NUTRITIONAL DEFICIENCY

Edited by **Pınar Erkekoglu** and **Belma Kocer-Gumusel**

Nutritional Deficiency
http://dx.doi.org/10.5772/61450
Edited by Pınar Erkekoglu and Belma Kocer-Gumusel

Contributors

Andreas Grabrucker, Ann Katrin Sauer, Simone Hagmeyer, Petra Bauer, Maria Augusta Naranjo Arcos, Naji Aljohani, Janet Schloss, Halima Nazar, Nidia Gomez, Verónica S. Biaggio, María Eugenia Ciminari, María V. Perez Chaca, Silvina M. Alvarez, Bernadeta Szewczyk, Anna Rafało, Gabriel Nowak, Magdalena Sowa-Kucma, Bartlomiej Pochwat, Ishag Adam, Gasim Gasim, Pablo Munoz

Published by InTech

Janeza Trdine 9, 51000 Rijeka, Croatia

Publishing Process Manager Iva Simcic

Technical Editor SPi Global

Cover InTech Design team

First published July, 2016

Printed in Croatia

Additional hard copies can be obtained from orders@intechopen.com

Nutritional Deficiency, Edited by Pınar Erkekoglu and Belma Kocer-Gumusel

p. cm.

Print ISBN 978-953-51-2437-5

Online ISBN 978-953-51-2438-2

For EU product safety concerns:
IN TECH d.o.o., Prolaz Marije Krucifikse Kozulić 3, 51000 Rijeka, Croatia,
info@intechopen.com or visit our website at intechopen.com.

Meet the editors

Pınar Erkekoglu graduated from TED Ankara College, Hacettepe University Faculty of Pharmacy (BS) and Anadolu University Public Relations School (BS). She is working as an associate professor in Hacettepe University, Faculty of Pharmacy, Department of Pharmaceutical Toxicology. She completed a part of her PhD studies on phthalates and oxidative stress in Grenoble, France, in Universite Joseph Fourier and CEA/INAC/LAN after receiving full scholarship from both the European Union Erasmus Scholarship Program and from Commissariat à l'énergie atomique et aux énergies alternatives (CEA). She received her PhD degree in 2009. She worked as a post-doc visiting associate in Massachusetts Institute of Technology (MIT) Biological Engineering Department during 2011–2012. She gave lectures as invited speaker in many congresses. Her main interests are the reproductive disorders, endocrine diseases, and oxidative stress induced by endocrine disrupting chemicals, particularly by phthalates, the protective role of selenium against oxidative stress, and the toxic effects of aromatic amines. Pınar Erkekoglu has published more than 100 papers in national and international peer-reviewed journals. She is a European Registered Toxicologist (ERT) since 2014.

Belma Kocer-Gumusel graduated from Hacettepe University Faculty of Pharmacy (BS). She is currently working as a full-time professor and serves as the department head in Hacettepe University, Faculty of Pharmacy, Department of Pharmaceutical Toxicology. She received her PhD degree in 1999. She worked as a postdoctoral fellow at the Department of Biochemistry, Laboratory of Oxidative Stress Biology in Grenoble Joseph Fourier University Faculty of Pharmacy and in CEA, the laboratory of Inorganic Chemistry and Nucleic Acid Lesions in 1999, 2000, and 2004. She gave lectures as invited speaker in many congresses. Her main interests are the reproductive disorders, endocrine diseases (particularly thyroid diseases), and oxidative stress induced by endocrine disrupting chemicals, particularly by phthalates, and the protective role of selenium against oxidative stress. Belma Kocer-Gumusel has published more than 100 papers in national and international peer-reviewed journals and has more than 700 citations. Dr. Kocer-Gumusel is a member of numerous national and international scientific societies including Turkish Society of Toxicology, EUROTOX, and IUTOX.

Contents

Preface

Nutritional deficiencies occur when the body does not receive enough nutrients it needs. Human body requires many different vitamins and minerals. Most of the nutrients are often referred to as micronutrients. As most of these vitamins and all of the minerals and trace elements are not produced naturally in the body, they must be obtained from the diet. Therefore, a healthy and well-balanced diet is essential for health and disease prevention. However, a big part of the world does not get enough food or healthy food due to economic or social reasons.

The amount of nutrients needed depends on several factors including age (infancy, childhood, adolescence, and elderly age), geographical region (i.e., soil and weather characteristics), and critical periods (e.g., pregnancy). Particularly in early periods of life, a well-balanced diet is very important for normal and healthy growth and development. Not getting enough vitamins, minerals, or trace elements in childhood can lead to growth retardation and specifically mental retardation. Iron deficiency is the most widespread nutritional deficiency worldwide. It can lead to anemia, which causes fatigue, weakness, and various other symptoms. Over 30% of the world's population suffers from this condition. On the other hand, zinc deficiency affects as much as 25% of the world's population, and zinc plays an essential role in numerous biochemical pathways. Vitamin D is essential for bone health. Its deficiency has become a worldwide issue in the elderly and remains common in children and adults for the last two decades. Vitamin D deficiency was suggested to be one of the underlying factors of diabetes, obesity, and muscle weakness.

Throughout this book, the readers will obtain information on different aspects of iron, zinc, and vitamin D deficiencies. Moreover, they will get information on the relationship between nutritional deficiency and cancer treatment. Besides, one chapter refers to disasters, food consumption, and nutritional status in developing countries, particularly addressing child nutrition in disasters. We believe that readers will gain scientific knowledge on nutritional deficiency, causes, and its consequences from this book. This book will also provide information on the effects of nutritional status in different diseases, particularly in cancer as well as in susceptible populations, such as pregnant women and elderly.

Assoc. Prof. Dr. Pınar Erkekoglu
Faculty of Pharmacy, Department of Pharmaceutical Toxicology
Hacettepe University, Ankara, Turkey
Prof. Dr. Belma Kocer-Gumusel
Faculty of Pharmacy, Department of Pharmaceutical Toxicology
Hacettepe University, Ankara, Turkey

Section 1

Zinc Deficiency

Chapter 1

Zinc Deficiency and Depression

Anna Rafalo, Magdalena Sowa-Kucma,
Bartlomiej Pochwat, Gabriel Nowak and
Bernadeta Szewczyk

Additional information is available at the end of the chapter

http://dx.doi.org/10.5772/63210

Abstract

Zinc deficiency has multiple effects, including neurological and somatic symptoms. Zinc deficiency can lead to depression, increased anxiety, irritability, emotional instability, and induced deficits in social behavior. Clinical studies have shown that low levels of zinc intake contributes to the symptoms of depression and patients suffering from depression have a lower serum zinc level. Also the animal studies have shown an important role of dietary zinc deficiency in the induction of depressive-like symptoms. Moreover, both preclinical and clinical studies have indicated the potential benefits of zinc supplementation as an adjunct to conventional antidepressant drugs or as a stand-alone intervention. This chapter focuses on the role of the zinc deficiency in the pathogenesis of depression, changes in animal behavior induced by dietary zinc restriction, the role of zinc supplementation in the treatment of depression, and the possible mechanisms involved in these relationships. Both clinical and preclinical studies related to these findings will be discussed.

Keywords: zinc, zinc deficiency, depression, zinc supplementation, biomarkers of depression

1. Introduction

Depression is a mental disorder associated with functional impairment, disability, morbidity, and mortality. Despite the extensive research, its pathophysiology is still poorly understood. Current pharmacotherapy, although effective is usually costly, requires time and has potential side effects [1, 2]. Thus, there is a strong need to investigate alternative prevention or treatment strategies. Both clinical and preclinical studies have indicated the important role of zinc in the pathophysiology and treatment of depression.

Early studies exploring the role of zinc in depression focused on demonstrating its antidepressant activity, and then, they tried to explain whether zinc enhances the effect of antidepressants and which mechanisms are involved in the antidepressant effects of zinc. Current studies, however, are mostly focused on examining whether serum zinc concentration might be a biomarker of depression or if experimentally induced zinc deficiency can be a new useful animal model of depression.

This review summarizes the most important data concerning the role of zinc and particularly zinc deficiency in the development or treatment of depression.

2. Dietary zinc intake and depressive symptoms (Table 1)

The first study indicating a relationship between dietary zinc deficiency and depression was performed by Amani et al. [3]. This study recruited 23 young women diagnosed with moderate-to-severe depression and 23 age matched healthy volunteers. All of the women involved in this study completed the food frequency questionnaire (FFQ) and a 24 h-food recall questionnaire to confirm the sources of zinc and daily zinc intakes in the diet. Analyses of the obtained results showed that both the daily zinc intake and the serum zinc concentration in the depressive group were lower than that found in the healthy women. Moreover, an inverse correlation between the depression scores and serum zinc concentration was found [3].

A few other papers published recently indicated that women seem to be more vulnerable to zinc deficiency than men. Maserejian et al. [4] examined the relationship between dietary zinc restriction and depressive symptoms in a large group of men and women. They analyzed cross-sectional, observational epidemiological data from the Boston Area Community Health (BACH) Survey. The final group of samples used in the analyses involved 2163 women and 1545 men from three racial/ethnic groups such as Hispanic, non-Hispanic black, and non-Hispanic white. To ascertain the diet from multiethnic populations, all of the participants completed the Food Frequency Questionnaire (FFQ). This questionnaire included also data on the vitamin and mineral supplements used. The depressive symptoms were assessed by the abridged Center for Epidemiologic Studies Depression scale (CES-D). Interestingly, it turned out that a zinc deficient diet influences the severity of depressive symptoms in women but not in men [4].

In the same year (2012), Jacka et al. [5] reported the data from a study on a large randomly selected, population-based groups of women (n = 1494, aged 20–94 years). Similarly, the aim of this study was to determine the relationship between the dietary intake of zinc as well as magnesium and folate and clinically determined depressive and anxiety symptoms. Habitual dietary intake was assessed using FFQ. The current depressive and anxiety depression were measured using the Structured Clinical Interview for DSM-IV-TR, non-patients edition, and the General Health Questionnaire-12 (GHQ-12) was used to assess the psychiatric symptoms. As reported in the results, the intake of both zinc and magnesium was inversely related to

GHQ-12 scores and each standard deviation increase in the intake of these nutrients was associated with a standard deviation decrease in GHQ-12 scores. The relationship between nutrient intakes and anxiety disorder was not significant [5].

Authors	Participants (women/men)	Measurement	Effect
Marcellini et al. [8]	(485/367)	FFQ, GDS, M-MSE, PSS	Low zinc intake associated with low zinc plasma level and psychological disorders
Amani et al. [3]	(308/0)	FFQ, serum Zn	Inverse relationship between depression symptoms and both dietary zinc deficiency and zinc serum level
Roy et al. [6]	(2030/0)	FFQ, CES-D	Zinc deficiency diet influences the severity of stress and symptoms of depression.
Jacka et al. [5]	(1046/0)	FFQ, GHQ-12,	Increase in zinc intake induced decrease in psychiatric symptoms but not anxiety
Maserejian et al. [4]	(2163/1545)	FFQ, CES-D	Zinc deficiency is associated with depression in women but not in men
Yary and Aazami [7]	(173/229)	FFQ, CES-D	Inverse relationship between dietary zinc intake and depression
Lehto et al. [10]	(0/2317)	CES-D	No effect of dietary zinc intake on the incidence of depression
Markie wicz-Zukowska et al. [9]	(52/48)	AMTS, GDS, SRH, ADL Serum Zn	Low serum zinc levels correlate with mental disorders and depression

ADL, Activities of Daily Living; AMTS, Abbreviated Mental Test Score; CES-D, Center for Epidemiologic Studies Depression scale; FFQ, Food Frequency Questionnaire; GDS-Geriatric Depression Scale; GHQ-General Health Questionnaire; M-MSE, Mini Mental State Examination; PSS, Perceived Stress Scale; SRH, Self-Rated Health.

* Analysis of cross-sectional, observational epidemiological data from the Boston Area Community Health (BACH) Survey.

Table 1. A summary of the influence of dietary zinc deficiency on the depressive symptoms.

Zinc intake as a causative factor in the induction of depressive symptoms was also considered in the context of prenatal depression. The study by Roy et al. [6] was conducted on pregnant women (2030 samples, recruited between 2002 and 2005 year) from the clinics in London, Ontario (Prenatal Health Project). The main aim of this study was to analyze the relationship between zinc intake, sociodemographic factors, and psychosocial stress, as well as the development of depressive symptoms. As in other studies, the zinc intake was assessed using FFQ and nutrient supplement data. The level of psychological stress was evaluated using a standardized composite score, and the depressive symptoms were measured using a CES-D scale. The results showed that low zinc intake, social disadvantage, and stress were correlated with higher CES-D scores, but higher zinc intake was found to buffer the impact of stress on the depressive symptoms during pregnancy [6].

Another study indicating that long-term zinc deficiency may induce the symptoms of depression was carried out on 402 postgraduate students with a mean age of about 32 years (229 men and 173 women) [7]. The CES-D questionnaire was used to measure the prevalence of depressive symptoms among participants, while the dietary intake of zinc among the postgraduate students was evaluated using the FFQ. The results of this study show an inverse relationship between dietary zinc intake and depression in postgraduate students [7].

The statistics showed that zinc deficiency is common not only among the children but also in the elderly. In the ZINCAGE project older people from five countries: Italy, Greece, Germany, France, and Poland were recruited to investigate the relationship between nutritional aspects in Northern and Southern European Countries, zinc status, and psychological dimensions (mood, perceived stress and cognitive functions) [8]. The FFQ was typically used to estimate the intake of zinc. Evaluation of mood, the perceived stress level, and cognitive function was meanwhile performed using the: "Mini Mental State Examination", Geriatric Depression Scale (GDS), and "Perceived Stress Scale." The study involved 853 people who were divided into four age groups. The first group included 359 people between 60–69 years old, the second 225 at the age of 70–74 years, the third 153 at the age of 75–79 years, and the last 116 age of 80–84 years. Eighty-two percentage of the total samples showed no cognitive decline, 72% (according to GDS) of samples showed no depression, and all of the samples had a normal perceived stress level. Interestingly, however, a relevant correlation between all of the psychological dimensions studied in the project and the plasma zinc values or nutritional assessment was found. This phenomenon was not correlated with the age, which indicates that a normal zinc diet can help maintain proper plasma zinc levels and a good psychological condition in the elderly [8].

These findings were confirmed in another screening study. A research group from Bialystok (Poland) [9] conducted a nearly 2-year (from October 2010 to May 2012) study among 100 nursing home residents (aged 60–102 years) to determine the zinc status among the older individuals in correlation to their mental and physical performance. The participants were subjected to the Abbreviated Mental Test Score (AMTS), Self-Rated Health (SRH), Geriatric Depression Scale (GDS), and Independence in Activities of Daily Living (ADL). Also anthropometric variables and fitness scores were evaluated. The serum zinc level was measured by flame atomic absorption spectrometry. Almost one-third of the participants had zinc deficiency. Cognitive functions were impaired in 45% of them, and 48% of the participants showed depressive symptoms. Moreover, it was found that elderly patients with normal cognitive function and without symptoms of depression had a higher serum zinc level than patients with mental disorders and depression [9].

One paper has so far showed no effect of dietary zinc intake on the incidence of depression [10]. This study was performed on 2317 Finnish men aged 42–61 years to estimate the correlation between dietary zinc intake and depression in a prospective setting in initially depression-free men during a 20-year follow-up. Nutrient intake was quantitatively evaluated by 4-day food record at the baseline. The severity of the depression was measured using the Human Population Laboratory Depression Scale. Participants who at the baseline had elevated depressive symptoms were excluded (n = 283) from the study. In this prospective setting, depression has been defined as a hospital discharge diagnosis of a unipolar depressive disorder [10]. Generally, ~3% of the participants received a hospital discharge diagnosis of depression during the 20-year follow-up, with the analysis adjusted for age, baseline depression severity, smoking, alcohol use, physical exercise, and the use of zinc dietary supplements not being associated with an increased depression risk in men [10]. However, as the author indicated, these observations may not be generalizable to women [10]. What is more, looking at the results of Maserejian et al. [4] and the other data described earlier, the present findings may simply confirm that zinc deficiency may play a greater role in the induction of depression in women.

3. Zinc level as a marker of depression (Table 2)

The biological markers (biomarkers) are defined as cellular, biochemical, or molecular alterations that are measurable in biological media such as human tissues, cells, or fluids. Biomarkers should be fixed and concern specific features of the disease which occurs permanently, regardless of the stage/phase of the illness (a trait marker) or changes, depending on the stage (a state marker) [11]. Appropriately chosen biomarkers would be more objective and, as such, would significantly enhance the traditional patient symptoms-based assessment of depression. Biomarkers of depression could be used to support the presence or absence of the disease, provide individualized treatment, monitor treatment progress (or indicate the risk of the drug resistance), and predict the onset of future disease or its relapse. A clinically useful biological marker should be characterized by a high level of sensitivity and specificity (preferably above 80%), and its determination should be easy to make and relatively cheap [12]. Some potential candidate markers of depression have been reported, for example, see [13], but none of them have yet been used in clinical practice.

Authors	Disorder	Results
Hansen et al. [11]	MDD	Reduced serum zinc level in depressed patients
Little et al. [14]	Mood disorder	Higher prevalence of zinc deficiency in outpatients with mood disorder
Amani et al. [3]; Maes et al. [15, 16]; McLoughlin and Hodge [18]; Siwek et al. [19]	MDD	Reduction in the blood zinc level in depressed patients

Authors	Disorder	Results
Maes et al. [15, 16]; Siwek et al. [19]; Wójcik et al. [20]	MDD	Negative correlation between severity of depressive symptoms and serum zinc level
Wójcik et al. [20]	–	Lower zinc level in women with antepartum depression
Wójcik et al. [20]; Brownlie and Legge [21]	–	Lower zinc level in women with postpartum depression
Roozbeh et al. [22]	MDD	Lower zinc level in depressed patients with end-stage renal disease undergoing hemodialysis
Marcellini et al. [8]	–	Lower zinc level in late life
Stanislawska et al. [23]	MDD	Negative correlation between postmenopausal women and the severity of depressive symptoms and serum zinc concentration
Crayton and Walsh [24]; Irmish et al. [25]; Nguyen et al. [26]	MDD	No differences in the zinc concentration between depressed and healthy people
Salustri et al. [27]	MDD	Increase in the peripheral blood level of zinc in the course of depression
Manser et al. [28]	MDD	Significant differences between serum zinc level in depressed male and women
McLoughlin and Hodge [18]	MDD	Antidepressant therapy normalized lower zinc level in depressed patients
Maes et al. [16]	MDD	The zinc levels in the treatment-resistant depressed patients were significantly lower than its concentration in the depressed patients and healthy volunteers
Maes et al. [16]; Siwek et al. [19]	MDD	Lack of normalization in zinc level after antidepressants treatment or its correlation with the severity of depression
Stanley and Wakwe [29]	MDD, BD, schizophrenia	Reduction in serum zinc level across all of the investigated groups
Gonzales-Estecha et al. [30]	BD	Significant increase in serum concentration in the manic and no changes in depressed phase when compared to the healthy volunteers
Siwek et al. [31]	BDI, BDII	Significantly decreased peripheral blood zinc level in depressive phase in BDI patients compared with healthy control no changes in BDII subgroups

MDD, major depressive disorder; BD, bipolar disorder.

Table 2. A summary of the clinical studies on the relationship between the zinc level and depression.

The first clinical report that suggested the potential role of zinc in the pathophysiology of depression and role of zinc as a marker of this disease was published in 1983 by Hansen and colleagues [11]. In this paper, the reduced serum zinc concentration in patients suffering from recurrent major depressive disorder and the negative correlation between the severity of depressive symptoms and the zinc level was described [11]. One of the preliminary studies also showed a significantly higher prevalence of zinc deficiency in outpatients with mood disorder than healthy controls [14]. The reduction of the blood zinc level in depressed patients has been confirmed in many other reports [3, 15–19]. Similarly, some studies have shown a negative correlation between the severity of depressive symptoms and serum zinc [15, 16, 19, 20]. The lower zinc level has also been observed in women with antepartum [20] and postpartum depression [20, 21], and in depressed patients with end-stage renal disease undergoing hemodialysis [22] or in late life [8]. Furthermore, in postmenopausal women, the negative correlation between the severity of depressive symptoms and the serum zinc concentration was noted [23]. To this date, only three studies have reported no differences in the zinc concentration between depressed and healthy people [24–26] and one report even shows an increase in the peripheral blood level of zinc in the course of depression [27]. It is also little data that show a significant differences between serum zinc in depressed male and women [28].

Support of the hypothesis that the zinc level might be a specific and sensitive marker of depression comes from the findings that the zinc level, lower in depressed patients, may be normalized to control levels after successful antidepressant therapy [18]. However, this effect was not observed in all of the studies. Maes et al. [16] showed that the zinc levels in the treatment-resistant depressed patients were significantly lower than its concentration in the remaining depressed patients and healthy volunteers. The next study by Maes et al. [17] performed on a larger group of patients, and a study by Siwek et al. [19] gave similar findings. A lack of normalization of the zinc level after antidepressants treatment or its correlation with the severity of depression was found. Based on this evidence, it has been suggested that a lower serum zinc level may be a sensitive (79%) and specific (93%) marker of drug resistance [16, 17].

To date, most studies on the role of zinc in the depression included patients diagnosed with major depressive disorder (MDD). Unfortunately, very little is known about the importance of zinc in the development (or treatment) of bipolar disorders (BD). So far, there are only three publications on this topic. In an early study, including patients diagnosed with MDD, BD and schizophrenia, a reduction in the serum zinc level across all of the investigated groups was observed [29]. A subsequent study, whose aim was to analyze the changes of trace elements exclusively in bipolar patients, indicated a significant increase in serum concentration in the manic and no changes in depressed phase when compared to the healthy volunteers [30]. Finally, a recent published study investigated in detail the issue concerning the concentration of zinc in bipolar disorder, with a special focus on the subtypes (BDI, BDII), and the phases and stages of the disease. It was found that a significantly decreased peripheral blood zinc level in the depressive phase in BDI patients (especially with stage 3 and 4) compared with the healthy control. However, both in remission and the manic phase, the zinc concentration was similar to that seen in the matched controls. In BDII subgroups, these alterations have not been noted [31].

Most of the clinical data come from studies of peripheral blood samples. Some researchers urged caution in interpreting such results. It is believed that the changes in the peripheral zinc concentration may not reflect zinc availability in the central nervous system, particularly in the brain. However, in healthy individuals, a clear correlation between the cerebrospinal fluid (CSF) and serum zinc level was demonstrated [32]. It has also been proved that zinc penetrates the blood–brain barrier [33]. So far, there are no studies to compare the concentrations of zinc in blood and/or CSF in depressed patient and healthy people. It is known that the hippocampal zinc concentration in suicide victims (where there is evidence for an association between suicidal behavior and depression) did not differ from that measured in sudden death controls [34, 35]. However, these data do not undermine the role of zinc in the pathophysiology of depression. For example, they may indicate that, in the course of depression, only subtle changes are present in the zinc level in the central tissues, which cannot be measured via the currently available analytical methods. Alternatively, these changes can occur in specific areas of the brain or even only in selected areas of certain brain structures that have not yet been investigated.

In conclusion, we can argue that the study of zinc in the context of depression and as a potential biomarker of the disease has great meaning and a future. Based on the presented research, we can argue that the serum zinc concentration may be a state marker of depression in treatment responder patients. Similarly, in drug-resistant depression, a reduced level of zinc may be a trait marker. While, in bipolar patients, an increased zinc level may be a state marker in mania. Unfortunately, due to the fact that alterations of zinc concentration are not specific only to depressive disorders, the measurement of this trace element in the blood of a patient cannot be a useful clinical marker of MDD or BD. It seems, however, that the determination of the serum zinc level could be in the future a component of multifactorial tests and assist diagnosis of the disease.

4. Zinc supplementation in the therapy of depression (Table 3)

There are still a limited number of clinical reports examining the effect of zinc supplementation on depressive symptoms. However, the available evidence suggests potential benefits of zinc supplementation as an adjunct to antidepressants therapy or as a stand-alone therapy for the prevention of depressive symptoms.

The first report indicating the beneficial effect of zinc supplementation in the therapy of depression was published in (2003) by Nowak et al. [36]. This placebo-controlled, double-blind pilot study was conducted in patients who fulfilled DSM-IV criteria for major (unipolar) depression. The recruited patients were divided into two groups: one receiving zinc supplementation (6 patients; 25 mg Zn/day) and the second (8 patients) a placebo. Both groups were treated with standard antidepressant therapy (clomipramine 125–150 mg; amitriptyline 125–150 mg; citalopram 20 mg; fluoxetine 20–40 mg). The efficacy of antidepressant therapy and the patient's status was evaluated before the treatment and 2, 6, and 12 weeks after it began using the Hamilton Depression Rating Scale (HDRS) and the Beck Depression Inventory (BDI). In this study, antidepressants significantly reduced HDRS scores by the 2nd week of treatment in both groups, and BDI scores at the 6th week in the zinc-treated group. Zinc supplementation significantly augmented this reduction after 6- and 12-week (HDRS) and at 12 week (BDI) of treatment when compared with the placebo. Although the observed effect of zinc was delayed, its potency was quite robust [36].

Authors	Participants (placebo/Zn group)	Supplementation (dose)	Duration (weeks)	Treatment (dose)	Measurement	Efficacy
Nowak et al. [36]	(8/6) MDD	Zinc hydroaspartate (25 mg Zn/day)	12	TCAs, SSRI	HDRS, BDI	Yes
Siwek et al. [37, 19]	(30/30) MDD	Zinc hydroaspartate (25 mg Zn/day)	12	TCAs	CGI, BDI, HDRS, MADRS	Yes
Ranjbar et al. [38, 39]	(17/20 or 21) MDD	Zinc sulfate (25 mg Zn/day)	12	SSRI	HDRS, BDI	Yes
Nguyen et al. [26]	396 (88/ 97/84/100) Healthy woman	Micronutrient supplements with zinc sulfate (~20 mg Zn/week or ~10 mg Zn/day)	12	No	CES-D, FFQ, serum Zn	No
Sawada and Yokoi [40]	(15/15) Healthy women	Multivitamins + zinc gluconate (7 mg Zn/day)	10	No	CMI, POMS	Yes
Di Girolamo et al. [43]	674 children	10 mg ZnO/day for 5 day/ week	6 months	No	Depression, anxiety, hyperactivity and conduct disorder	No
Maserejian et al. [4]	(2163 women/ 1545 men)	Multivitamins (0.1–15 mg Zn/day) or zinc supplements (>15 mg Zn/day)	NA	No or SSRI, SNRI, serotonin modulators	FFQ, CES-D	Yes

BDI, Beck Depression Inventory; CES-D, Center for Epidemiologic Studies Depression scale; CGI, Clinical Global Impression; CMI, Cornell Medical Index (somatic symptoms and mental symptoms); FFQ, Food Frequency Questionnaire; HDRS, Hamilton Depression Rating Scale; MADRS, Montgomery–Asberg Depression Rating Scale; POMS, Profile of Mood State (depression dejection and anger hostility score); NA, not applicable.

* The final group of patients.

** Analysis of cross-sectional, observational epidemiological data from the Boston Area Community Health (BACH) Survey.

Table 3. Summary of studies on the efficacy of zinc supplementation in the treatment or prevention of depressive symptoms.

The second report of the benefit of zinc supplementation in antidepressant therapy was published by Siwek et al. [37]. This group performed a placebo-controlled, double-blind study of zinc supplementation in imipramine therapy in sixty, 18–55-year old, unipolar depressed patients fulfilling the DSM-IV criteria for major depression but without psychotic symptoms. The participants were randomized into two groups treated with imipramine (approximately

140 mg/day) and receiving once daily either a placebo or zinc supplementation (25 mg Zn/day) for 12 weeks. In this study, the BDI, HDRS, CGI, and MADRS scales were used. Analyses of the results obtained from this study showed no differences in CGI, BDI, HADRS, and MADRS scores between zinc-supplemented and placebo-supplemented antidepressant treated non-resistant patients. However, they indicated that zinc supplementation reduced depression scores and augmented the efficacy and speed of the onset of response to antidepressant treatment in the patients previously non-responsive to antidepressant therapy [19, 37].

A study by Ranjbar et al. [38, 39] presented a double-blind randomized clinical trial on 39 patients (aged 18–55 years) diagnosed with MDD. The participants of this study were randomly assigned to groups receiving zinc supplementation (25 mg Zn/day) or a placebo. The patients from both groups received SSRIs (citalopram 20–60 mg/day or fluoxetine 20–60 mg/day) for 12 weeks. The severity of depression was measured using the BDI [38] and HDRS [39] at the baseline and after 6 and 12 weeks of treatment. At the end of the study, the BDI and HDRS scores were significantly lower in the SSRI and zinc-treated group than the group receiving the SSRI and the placebo.

The report published by Sawada and Yokoi [40] showed in turn that young women taking multivitamins and zinc supplements exhibited a significant reduction in depression and anxiety symptoms than women taking only multivitamins. This randomized, double-blind, placebo-controlled study was performed among 30 women in the aged of 18–21 years. The subjects were randomly assigned to receive multivitamin capsules or multivitamins and zinc once daily for 10 weeks. The Cornell Medical Index was used to evaluate somatic symptoms (A-L score) and mental symptoms (mood and feelings, including anxiety, sensitivity, anger and tension, M-R score). Additionally to ascertain the mood state during the previous week, the Profile of Mood State (POMS) questionnaire was used. Analyses of the data highlighted that women taking multivitamins and zinc showed a significant decrease in anger-hostility and depression-dejection scores (on the POMS). No changes were found in the M-R (mental symptoms) on the CMI between women taking multivitamins and zinc and multivitamins only. Despite this, as the authors themselves make clear, these findings are only preliminary and do not suggest that zinc supplementation may be effective in reducing anger and depression [40].

Another paper published in 2012 by Maserejian et al. [4] (the study design and population was described in detail in the Section 2) revealed that dietary, supplemental, and total zinc were significantly associated with the presence of depressive symptoms but only in women and not in men. Moreover, there was a statistically significant interaction between the total zinc intake and use of SSRIs in the development of depressive symptoms. No statistically significant interactions were observed for use of SNRIs, TCAs, or antipsychotics [4].

In the same year (2012), Sandstead [41] published the results from six randomized controlled comparative treatment experiments performed on Chinese and Mexican-American low-income children (6–9 years); middle-income US premenopausal women; middle-income US adolescents, and middle-income US men. These findings illustrated that subclinical zinc deficiency changes the brain function and that zinc and micronutrient treatment improves altered brain functions [41]. Recently, the beneficial effect of zinc monotherapy (30 mg Zn,

12 weeks) in the relief of depressive symptoms in overweight or obese subjects was also reported [42].

Two studies have so far shown no effect of zinc supplementation on the improvement of depressive symptoms [26, 43]. These studies, however, differ significantly from the one previously described with respect to both the participants and the length and quality of applications. The first study by DiGirolamo et al. [43] examined the effect of zinc supplementation on the mental health of school-age children in Guatemala. Zinc as a ZnO (10 mg/day for 5 days/week) was applied for 6 months. Outcome measures at the end of the study included internalizing problems such as depression or anxiety and externalizing (hyperactivity and conduct disorder) problem behaviors. Generally, no difference in mental health outcomes between the zinc and placebo groups was found, although increases in serum zinc concentrations were associated with decreases in depression and anxiety among the children who were at risk of zinc deficiency [43]. The second study of Nguyen et al. [26] investigated the impact of combinations of micronutrient supplements on symptoms of depression rather than the effect of zinc supplementation as a stand-alone therapy.

Because of these methodological limitations in the existing studies, further well-designed, adequately powered research is required.

The data described earlier are clinical verification of the results obtained in preclinical studies. The beneficial effects of zinc treatment have been in fact reported in several preclinical test and models of depression. Zinc administration induced an antidepressant-like effect in the forced swim test (FST) both in mice and rats and in the tail suspension test (TST) in mice [44–49]. Zinc was also active in different models of depression such as: (a) the olfactory bulbectomy (OB) (a reduction in the number of trials in the passive-avoidance test and a decreased OB-induced hyperactivity in rats) [48]; (b) the chronic mild stress (CMS) model of depression (zinc reversed the CMS-induced reduction in the consumption of sucrose in rats) [50]; and (c) chronic unpredictable stress (CUS) (zinc treatment prevented deficits in the fighting behavior of chronically stressed rats) [51]. Moreover, zinc has been found to intensify the effects of standard antidepressants (imipramine, fluoxetine, paroxetine, bupropion or citalopram) in the FST, the TST, and CUS [44, 49, 51–53].

5. Experimental zinc deficiency as an animal model of depression

Analysis of the clinical data makes it possible to address a very important question concerning the relationship between zinc deficiency and depression, namely to establish whether zinc deficiency is the cause or an important risk factor for depression or if zinc deficiency is a consequence of pathological processes underlying depression [54, 55]. In order to answer this question, several animal studies have been conducted. It is well established that the subjection of animals to procedures such as chronic stress, removal of olfactory bulbs, and several other leads to observed, depressive-like behaviors. In principle, these changes in animal behavior in some extent should correspond with human behaviors observed in depressive patients, that is, olfactory bulbectomy in rodents serves as a model of agitated depression, and chronic mild stress modulates anhedonia in laboratory animals. Several tests are used to evaluate the pro-depressive effects of a particular procedure, with the most popular being the open-field test, social interaction, FST, TST, and sucrose intake test.

Several data published recently indicated that experimentally induced zinc deficiency may be one of the procedures used to modulate depressive symptoms in both mice and rats. According to the data in the so-called "experimentally induced zinc deficiency" model two factors play an important role: the amount of zinc in the feed and the time of exposure to zinc restriction. Whittle et al. [56] showed that feed containing 40% of zinc in the daily requirement (~12 mgZn/kg) is sufficient to produce depressive-like behavior in mice, which is observed as increased immobility time in the TST and FST [56]. In this study, mice were treated with zinc deficient diet for 7 weeks. Mlyniec et al. [57, 58] showed that 4 or 10 weeks of zinc deficiency (0.2 mgZn/kg) in mice induced depressive-like behavior observed in TST and FST. Interestingly, the same studies showed that mice subjected to a zinc deficient diet for 2 weeks displayed antidepressant-like activity in the TST and FST [57, 58]. The explanation of these differences requires further studies. The more, that in contrast to the results obtained by Mlyniec et al. [58], 2 weeks exposure to zinc deficiency (0.37 mgZn/kg) induced the depressive-like behavior in young rats, measured as and increased immobility time in the FST [59, 60]. In studies conducted by Tassabehji et al. [61], 3 weeks of zinc deficiency (10 mgZn/kg) led to increased immobility time in the FST and decreased sucrose intake in adult rats. Similar results were obtained by Doboszewska et al. [62] who showed that 4 or 6 weeks consumption of feed low in zinc (3 mg Zn/kg) caused the decreased intake of sucrose solution and increased immobility time in the FST. Additionally, 4 or 6 weeks zinc deficiency significantly reduced social interaction in adult rats [62].

As shown above, the behavioral disturbances induced by zinc deficiency overlap with depressive-like behaviors induced by well-known experimental procedures. Additionally, some of these procedures, such as stress, are causally related to a lower serum zinc level in rats. However, the physiological role of zinc is very complex and involves the activity of zinc on many receptors and enzymes, with some of these molecular events possibly being considered as key factors engaging in depression or depressive-like behaviors. Zinc is an inhibitor of glutamate N-methyl-D-aspartate receptors (NMDAR), therefore zinc released with glutamate from glutamate terminals determines the correct functioning of the glutamate system [63, 64]. In pathological states, when the glutamate concentration is radically increased in the synaptic cleft, the overstimulation of NMDAR may lead to atrophy and neural cell death. The overstimulation of NMDAR leading to atrophy cell death is named excitotoxicity [65, 66]. The role of glutamate transmission in behavioral abnormalities induced by zinc deficiency was indicated by Doboszewska et al. [62]. They showed that zinc deprivation led to a significantly enhanced expression of GluN1, GluN2A, and GluN2B subunits of NMDAR in the hippocampus and GluN2B subunit in the prefrontal cortex (PFC) [62, 67]. In the same study, zinc deficiency radically decreased the level of brain-derived neurotrophic factor (BDNF), whose concentration in the physiological range is a crucial factor for normal neurotransmission and the survival of neurons. A decreased level of BDNF has been noted in both clinical and preclinical studies, respectively, in depressive patients and animals subjected to procedures inducing depressive-like behaviors [68].

One of the factors modulating the glutamate transmission is glucocorticoids [69]. The enhanced level of glucocorticoids may potentiate glutamate enhanced transmission in the hippocampus, leading to neuronal death [70]. Only 1 week of zinc deprivation is enough to decrease the serum zinc level in rats with a concomitant increase level of glucocorticoids serum levels [69–71]. These data indicate on hyperactivation of the hypothalamic-pituitary-adrenal axis (HPA-axis) in a zinc deficient condition. However, the causative relationship between zinc deficiency and HPA-axis overstimulation is unknown. The enhanced activation of the HPA-axis disrupts hippocampal function which is a brain structure sensitive to a higher level of glucocorticoids and involved in the development and expression of depression [72]. Moreover, a similar effect of the overstimulation of the HPA-axis on hippocampal function has been observed under stressful conditions [72].

6. Conclusions

The importance of zinc as the life-threatening factor for humans was first described in 1963 by Prasad et al. [73, 74]. Zinc deficiency still remains a substantial global health problem. Statistics show that two billion people worldwide are not getting enough zinc via their diet. Zinc deficiency is accountable for physical and intellectual retardation, preventing children from developing to their full potential. Marginal zinc deficiency is also evident in older people. A recent study showed a significant relationship between zinc deficiency and cognitive impairment, increased susceptibility to stress and something that should be emphasized, to depressive symptoms. Taking into account that depression is the most serious mental illness associated with decreased productivity and quality of life and well-being, evidence about the causative role of zinc deficiency in the development of this disease is very important.

Recent clinical research, indicated that the serum zinc level in patients was significantly lower than in the control group. Other studies have also shown that low serum zinc levels are normalized during treatment with antidepressants [54]. This suggests that the measurement of the serum zinc level could be in the future a component of multifactorial tests and assist diagnosis of the disease. This is of particular importance, especially now, when effective markers of the disease are needed. This problem also applies to mental illnesses. The next significant aspect is the pharmacotherapy of depression. Current treatments for depression are costly, have potential side effects, and require time and commitment. However, most alarming is the fact that antidepressants are effective only in 60% of patients. Treatments with zinc in laboratory animals have antidepressant effects, and zinc supplementation enhances the effectiveness of antidepressants in animal tests and models of depression. Also, clinical studies examined the effects of zinc supplementation as an adjunct to antidepressant drug therapy. Both preclinical and clinical studies indicated that zinc supplementation could be an adjunct increasing the effectiveness of tricyclic antidepressants but especially, selective serotonin reuptake inhibitors rather than selective noradrenaline reuptake inhibitors [38]. These results, therefore, suggest that zinc supplementation may be effective but only for drugs with a specific mechanism of action. To ensure the benefits of zinc supplementation as an adjunct to conventional antidepressants or as an intervention in the prevention of depression or as a marker of depression more, high-quality trials are needed.

Acknowledgements

This study was supported by funds from the Statutory Activity of the Institute of Pharmacology, Polish Academy of Sciences; Krakow, Poland. (AR) is a participant of the Ph.D. program from the Jagiellonian University, Krakow, Poland.

Author details

Anna Rafalo[1,2], Magdalena Sowa-Kucma[1], Bartlomiej Pochwat[1], Gabriel Nowak[1,3] and Bernadeta Szewczyk[1*]

*Address all correspondence to: szewczyk@if-pan.krakow.pl

1 Department of Neurobiology, Institute of Pharmacology, Polish Academy of Sciences, Krakow, Poland

2 Institute of Zoology, Jagiellonian University, Krakow, Poland

3 Faculty of Pharmacy, Jagiellonian University Medical College, Kraków, Poland

References

[1] Hansen R, Gaynes B, Thieda P, Gartlehner G, Deveaugh-Geiss A, Krebs E, Lohr K: Meta-analysis of major depressive disorder relapse and recurrence with second-generation antidepressants. Psychiatr Serv 2008;59:1121–1130.

[2] Nemeroff CB: The burden of severe depression: a review of diagnostic challenges and treatment alternatives. J Psychiatr Res 2007;41:189–206.

[3] Amani R, Saeidi S, Nazari Z, Nematpour S: Correlation between dietary zinc intakes and its serum levels with depression scales in young female students. Biol Trace Elem Res 2010;137:150–158.

[4] Maserejian NN, Hall SA, McKinlay JB: Low dietary or supplemental zinc is associated with depression symptoms among women, but not men, in a population-based epidemiological survey. J Affect Disord 2012;136:781–788.

[5] Jacka FN, Maes M, Pasco JA, Williams LJ, Berk M: Nutrient intakes and the common mental disorders in women. J Affect Disord 2012;141:79–85.

[6] Roy A, Evers SE, Avison WR, Campbell MK: Higher zinc intake buffers the impact of stress on depressive symptoms in pregnancy. Nutr Res 2010;30:695–704.

[7] Yary T, Aazami S: Dietary intake of zinc was inversely associated with depression. Biol Trace Elem Res 2012;145:286–290.

[8] Marcellini F, Giuli C, Papa R, Gagliardi C, Dedoussis G, Herbein G, Fulop T, Monti D, Rink L, Jajte J, Mocchegiani E: Zinc status, psychological and nutritional assessment in old people recruited in five European countries: Zincage study. Biogerontology 2006;7:339–345.

[9] Markiewicz-Zukowska R, Gutowska A, Borawska MH: Serum zinc concentrations correlate with mental and physical status of nursing home residents. Plos One 2015;10:e0117257.

[10] Lehto SM, Ruusunen A, Tolmunen T, Voutilainen S, Tuomainen TP, Kauhanen J: Dietary zinc intake and the risk of depression in middle-aged men: a 20-year prospective follow-up study. J Affect Disord 2013;150:682–685.

[11] Hansen CR, Jr., Malecha M, Mackenzie TB, Kroll J: Copper and zinc deficiencies in association with depression and neurological findings. Biol Psychiatry 1983;18:395–401.

[12] Kalia M, Costa E Silva: Biomarkers of psychiatric diseases: current status and future prospects. Metabolism 2015;64:S11–S15.

[13] Huang TL, Lin CC: Advances in biomarkers of major depressive disorder. Adv Clin Chem 2015;68:177–204.

[14] Little KY, Castellanos X, Humphries LL, Austin J: Altered zinc metabolism in mood disorder patients. Biol Psychiatry 1989;26:646–648.

[15] Maes M, D'Haese PC, Scharpe S, D'Hondt P, Cosyns P, De Broe ME: Hypozincemia in depression. J Affect Disord 1994;31:135–140.

[16] Maes M, Bosmans E, De JR, Kenis G, Vandoolaeghe E, Neels H: Increased serum IL-6 and IL-1 receptor antagonist concentrations in major depression and treatment resistant depression. Cytokine 1997;9:853–858.

[17] Maes M, De VN, Demedts P, Wauters A, Neels H: Lower serum zinc in major depression in relation to changes in serum acute phase proteins. J Affect Disord 1999;56:189–194.

[18] McLoughlin IJ, Hodge JS: Zinc in depressive disorder. Acta Psychiatr Scand 1990;82:451–453.

[19] Siwek M, Dudek D, Schlegel-Zawadzka M, Morawska A, Piekoszewski W, Opoka W, Zieba A, Pilc A, Popik P, Nowak G: Serum zinc level in depressed patients during zinc supplementation of imipramine treatment. J Affect Disord 2010;126:447–452.

[20] Wójcik J, Dudek D, Schlegel-Zawadzka M, Grabowska M, Marcinek A, Florek E, Piekoszewski W, Nowak RJ, Opoka W, Nowak G: Antepartum/postpartum depressive symptoms and serum zinc and magnesium levels. Pharmacol Rep 2006;58:571–576.

[21] Brownlie BE, Legge HM: Thyrotropin results in euthyroid patients with a past history of hyperthyroidism. Acta Endocrinol (Copenh) 1990;122:623–627.

[22] Roozbeh J, Sharifian M, Ghanizadeh A, Sahraian A, Sagheb MM, Shabani S, Hamidian JA, Kashfi M, Afshariani R: Association of zinc deficiency and depression in the patients with end-stage renal disease on hemodialysis. J Ren Nutr 2011;21:184–187.

[23] Stanislawska M, Szkup-Jablonska M, Jurczak A, Wieder-Huszla S, Samochowiec A, Jasiewicz A, Nocen I, Augustyniuk K, Brodowska A, Karakiewicz B, Chlubek D, Grochans E: The severity of depressive symptoms vs. serum Mg and Zn levels in postmenopausal women. Biol Trace Elem Res 2014;157:30–35.

[24] Crayton JW, Walsh WJ: Elevated serum copper levels in women with a history of post-partum depression. J Trace Elem Med Biol 2007;21:17–21.

[25] Irmisch G, Schlaefke D, Richter J: Zinc and fatty acids in depression. Neurochem Res 2010;35:1376–1383.

[26] Nguyen PH, Grajeda R, Melgar P, Marcinkevage J, DiGirolamo AM, Flores R, Martorell R: Micronutrient supplementation may reduce symptoms of depression in Guatemalan women. Arch Latinoam Nutr 2009;59:278–286.

[27] Salustri C, Squitti R, Zappasodi F, Ventriglia M, Bevacqua MG, Fontana M, Tecchio F: Oxidative stress and brain glutamate-mediated excitability in depressed patients. J Affect Disord 2010;127:321–325.

[28] Manser WW, Khan MA, Hasan KZ: Trace element studies on Karachi population. Part IV: blood copper, zinc, magnesium and lead levels in psychiatric patients with depression, mental retardation and seizure disorders. J Pak Med Assoc 1989;39:269–274.

[29] Stanley PC, Wakwe VC: Toxic trace metals in the mentally ill patients. Niger Postgrad Med J 2002;9:199–204.

[30] Gonzalez-Estecha M, Trasobares EM, Tajima K, Cano S, Fernandez C, Lopez JL, Unzeta B, Arroyo M, Fuentenebro F: Trace elements in bipolar disorder. J Trace Elem Med Biol 2011;25 Suppl 1:S78–S83.

[31] Siwek M, Sowa-Kucma M, Styczen K, Szewczyk B, Reczynski W, Misztak P, Topor-Madry R, Nowak G, Dudek D, Rybakowski JK: Decreased serum zinc concentration during depressive episode in patients with bipolar disorder. J Affect Disord 2016;190:272–277.

[32] Palm R, Sjostrom R, Hallmans G: Optimized atomic absorption spectrophotometry of zinc in cerebrospinal fluid. Clin Chem 1983;29:486–491.

[33] Pullen RG, Franklin PA, Hall GH: 65Zn uptake from blood into brain in the rat. J Neurochem 1991;56:485–489.

[34] Nowak G, Szewczyk B, Sadlik K, Piekoszewski W, Trela F, Florek E, Pilc A: Reduced potency of zinc to interact with NMDA receptors in hippocampal tissue of suicide victims. Pol J Pharmacol 2003;55:455–459.

[35] Sowa-Kucma M, Szewczyk B, Sadlik K, Piekoszewski W, Trela F, Opoka W, Poleszak E, Pilc A, Nowak G: Zinc, magnesium and NMDA receptor alterations in the hippocampus of suicide victims. J Affect Disord 2013;151:924–931.

[36] Nowak G, Siwek M, Dudek D, Zieba A, Pilc A: Effect of zinc supplementation on antidepressant therapy in unipolar depression: a preliminary placebo-controlled study. Pol J Pharmacol 2003;55:1143–1147.

[37] Siwek M, Dudek D, Paul IA, Sowa-Kucma M, Zieba A, Popik P, Pilc A, Nowak G: Zinc supplementation augments efficacy of imipramine in treatment resistant patients: a double blind, placebo-controlled study. J Affect Disord 2009;118:187–195.

[38] Ranjbar E, Kasaei MS, Mohammad-Shirazi M, Nasrollahzadeh J, Rashidkhani B, Shams J, Mostafavi SA, Mohammadi MR: Effects of zinc supplementation in patients with major depression: a randomized clinical trial. Iran J Psychiatry 2013;8:73–79.

[39] Ranjbar E, Shams J, Sabetkasaei M, Shirazi M, Rashidkhani B, Mostafavi A, Bornak E, Nasrollahzadeh J: Effects of zinc supplementation on efficacy of antidepressant therapy, inflammatory cytokines, and brain-derived neurotrophic factor in patients with major depression. Nutr Neurosci 2014;17:65–71.

[40] Sawada T, Yokoi K: Effect of zinc supplementation on mood states in young women: a pilot study. Eur J Clin Nutr 2010;64:331–333.

[41] Sandstead HH: Subclinical zinc deficiency impairs human brain function. J Trace Elem Med Biol 2012;26:70–73.

[42] Solati Z, Jazayeri S, Tehrani-Doost M, Mahmoodianfard S, Gohari MR: Zinc monotherapy increases serum brain-derived neurotrophic factor (BDNF) levels and decreases depressive symptoms in overweight or obese subjects: a double-blind, randomized, placebo-controlled trial. Nutr Neurosci 2015;18:162–168.

[43] DiGirolamo AM, Ramirez-Zea M, Wang M, Flores-Ayala R, Martorell R, Neufeld LM, Ramakrishnan U, Sellen D, Black MM, Stein AD: Randomized trial of the effect of zinc supplementation on the mental health of school-age children in Guatemala. Am J Clin Nutr 2010;92:1241–1250.

[44] Cunha MP, Machado DG, Bettio LE, Capra JC, Rodrigues AL: Interaction of zinc with antidepressants in the tail suspension test. Prog Neuropsychopharmacol Biol Psychiatry 2008;32:1913–1920.

[45] Franco JL, Posser T, Brocardo PS, Trevisan R, Uliano-Silva M, Gabilan NH, Santos AR, Leal RB, Rodrigues AL, Farina M, Dafre AL: Involvement of glutathione, ERK1/2 phosphorylation and BDNF expression in the antidepressant-like effect of zinc in rats. Behav Brain Res 2008;188:316–323.

[46] Kroczka B, Zieba A, Dudek D, Pilc A, Nowak G: Zinc exhibits an antidepressant-like effect in the forced swimming test in mice. Pol J Pharmacol 2000;52:403–406.

[47] Kroczka B, Branski P, Palucha A, Pilc A, Nowak G: Antidepressant-like properties of zinc in rodent forced swim test. Brain Res Bull 2001;55:297–300.

[48] Nowak G, Szewczyk B, Wieronska JM, Branski P, Palucha A, Pilc A, Sadlik K, Piekoszewski W: Antidepressant-like effects of acute and chronic treatment with zinc in forced swim test and olfactory bulbectomy model in rats. Brain Res Bull 2003;61:159–164.

[49] Rosa AO, Lin J, Calixto JB, Santos AR, Rodrigues AL: Involvement of NMDA receptors and L-arginine-nitric oxide pathway in the antidepressant-like effects of zinc in mice. Behav Brain Res 2003;144:87–93.

[50] Sowa-Kucma M, Legutko B, Szewczyk B, Novak K, Znojek P, Poleszak E, Papp M, Pilc A, Nowak G: Antidepressant-like activity of zinc: further behavioral and molecular evidence. J Neural Transm 2008;115:1621–1628.

[51] Cieslik K, Klenk-Majewska B, Danilczuk Z, Wrobel A, Lupina T, Ossowska G: Influence of zinc supplementation on imipramine effect in a chronic unpredictable stress (CUS) model in rats. Pharmacol Rep 2007;59:46–52.

[52] Szewczyk B, Branski P, Wieronska JM, Palucha A, Pilc A, Nowak G: Interaction of zinc with antidepressants in the forced swimming test in mice. Pol J Pharmacol 2002;54:681–685.

[53] Szewczyk B, Poleszak E, Wlaz P, Wrobel A, Blicharska E, Cichy A, Dybala M, Siwek A, Pomierny-Chamiolo L, Piotrowska A, Branski P, Pilc A, Nowak G: The involvement of serotonergic system in the antidepressant effect of zinc in the forced swim test. Prog Neuropsychopharmacol Biol Psychiatry 2009;33:323–329.

[54] Siwek M, Szewczyk B, Dudek D, Styczen K, Sowa-Kucma M, Mlyniec K, Siwek A, Witkowski L, Pochwat B, Nowak G: Zinc as a marker of affective disorders. Pharmacol Rep 2013;65:1512–1518.

[55] Szewczyk B, Poleszak E, Sowa-Kucma M, Siwek M, Dudek D, Ryszewska-Pokrasniewicz B, Radziwon-Zaleska M, Opoka W, Czekaj J, Pilc A, Nowak G: Antidepressant activity of zinc and magnesium in view of the current hypotheses of antidepressant action. Pharmacol Rep 2008;60:588–589.

[56] Whittle N, Lubec G, Singewald N: Zinc deficiency induces enhanced depression-like behaviour and altered limbic activation reversed by antidepressant treatment in mice. Amino Acids 2009;36:147–158.

[57] Mlyniec K, Nowak G: Zinc deficiency induces behavioral alterations in the tail suspension test in mice. Effect of antidepressants. Pharmacol Rep 2012;64:249–255.

[58] Mlyniec K, Davies CL, Budziszewska B, Opoka W, Reczynski W, Sowa-Kucma M, Doboszewska U, Pilc A, Nowak G: Time course of zinc deprivation-induced alterations of mice behavior in the forced swim test. Pharmacol Rep 2012;64:567–575.

[59] Tamano H, Kan F, Kawamura M, Oku N, Takeda A: Behavior in the forced swim test and neurochemical changes in the hippocampus in young rats after 2-week zinc deprivation. Neurochem Int 2009;55:536–541.

[60] Watanabe M, Tamano H, Kikuchi T, Takeda A: Susceptibility to stress in young rats after 2-week zinc deprivation. Neurochem Int 2010;56:410–416.

[61] Tassabehji NM, Corniola RS, Alshingiti A, Levenson CW: Zinc deficiency induces depression-like symptoms in adult rats. Physiol Behav 2008;95:365–369.

[62] Doboszewska U, Sowa-Kucma M, Mlyniec K, Pochwat B, Holuj M, Ostachowicz B, Pilc A, Nowak G, Szewczyk B: Zinc deficiency in rats is associated with up-regulation of hippocampal NMDA receptor. Prog Neuropsychopharmacol Biol Psychiatry 2015;56:254–263.

[63] Paoletti P, Ascher P, Neyton J: High-affinity zinc inhibition of NMDA NR1–NR2A receptors. J Neurosci 1997;17:5711–5725.

[64] Vergnano AM, Rebola N, Savtchenko LP, Pinheiro PS, Casado M, Kieffer BL, Rusakov DA, Mulle C, Paoletti P: Zinc dynamics and action at excitatory synapses. Neuron 2014;82:1101–1114.

[65] Hardingham GE, Bading H: Synaptic versus extrasynaptic NMDA receptor signalling: implications for neurodegenerative disorders. Nat Rev Neurosci 2010;11:682–696.

[66] Parsons MP, Raymond LA: Extrasynaptic NMDA receptor involvement in central nervous system disorders. Neuron 2014;82:279–293.

[67] Doboszewska U, Szewczyk B, Sowa-Kucma M, Mlyniec K, Rafalo A, Ostachowicz B, Lankosz M, Nowak G: Antidepressant activity of fluoxetine in the zinc deficiency model in rats involves the NMDA receptor complex. Behav Brain Res 2015;287:323–330.

[68] Duman RS, Voleti B: Signaling pathways underlying the pathophysiology and treatment of depression: novel mechanisms for rapid-acting agents. Trends Neurosci 2012;35:47–56.

[69] Takeda A, Tamano H, Ogawa T, Takada S, Ando M, Oku N, Watanabe M: Significance of serum glucocorticoid and chelatable zinc in depression and cognition in zinc deficiency. Behav Brain Res 2012;226:259–264.

[70] Takeda A, Tamano H: Insight into zinc signaling from dietary zinc deficiency. Brain Res Rev 2009;62:33–44.

[71] Takeda A, Tamano H, Kan F, Hanajima T, Yamada K, Oku N: Enhancement of social isolation-induced aggressive behavior of young mice by zinc deficiency. Life Sci 2008;82:909–914.

[72] Nestler EJ, Barrot M, DiLeone RJ, Eisch AJ, Gold SJ, Monteggia LM: Neurobiology of depression. Neuron 2002;34:13–25.

[73] Halsted JA, Prasad AS: Zinc deficiency in man. Isr Med J 1963;22:307–315.

[74] Prasad AS, Schulert AR, Miale A, Jr., Farid Z, Sandstead HH: Zinc and iron deficiencies in male subjects with dwarfism and hypogonadism but without ancylostomiasis, schistosomiasis or severe anemia. Am J Clin Nutr 1963;12:437–444.

Chapter 2

Zinc Deficiency

Ann Katrin Sauer, Simone Hagmeyer and
Andreas M. Grabrucker

Additional information is available at the end of the chapter

http://dx.doi.org/10.5772/63203

Abstract

Zinc is an essential trace element for humans and plays a critical role both as a structural component of proteins and as a cofactor in about 300 enzymes. Zinc deficiency was, for example, reported to affect the immune response and the endocrine system and to induce and modify brain disorders. Besides hereditary zinc deficiency, zinc deficiency – at least in mild forms – is nowadays a very abundant health issue. Today, an estimated 20% of the population worldwide is at risk of developing zinc deficiency with a high number also in industrialized countries. The major risk factors to develop zinc deficiency in industrialized nations are aging and pregnancy. Mechanistic and behavioral studies on the effects of zinc deficiency have mainly been performed using animal models. However, in combination with the few studies on human subjects, a picture emerges that shows importance of adequate nutritional zinc supply for many processes in the body. Especially the immune system and brain development and function seem to be highly sensitive to zinc deficiency. Here, we provide an overview on the effects of zinc deficiency on different organ systems, biological processes, and the associations of zinc deficiency with pathologies observed in humans and animal models.

Keywords: Zn, immune system, brain, homeostasis, synapse, biometal, trace metal, zincergic

1. Introduction

In 1933, zinc was reported for the first time to be essential for the growth of rats. Thirty years later, the first studies in human subjects from the Middle East showed that this was also true for humans [1,2]. To date, many studies have been performed investigating the influence of zinc deficiency on human well-being and mental performance.

While zinc deficiency is commonly caused by dietary factors, several inherited defects of zinc deficiency have been identified. Among them, *Acrodermatitis enteropathica* (AE) is the most common form of inherited zinc deficiency in humans [3]. Inherited AE is an autosomal recessive disorder where in many of the cases mutations in hZIP4 (a member of the SLC39 gene family encoding a membrane-bound zinc transporter) are found [4]. Mutations in other members of this family or in different zinc homeostasis genes may account for other cases of AE in the absence of hZIP4 mutations [5]. Clinically, AE is characterized by impaired intestinal zinc absorption, resulting in a triad of symptoms: dermatitis, alopecia, and gastrointestinal (GI) problems such as intractable diarrhea. However, neuropsychological disturbances such as mental depression, irritability, loss of appetite, behavioral problems, and reduced immune function frequently occur.

The body of an adult human (70 kg) contains about 2–3 g of zinc, which is absorbed from our dietary sources in the proximal small intestine, either the distal duodenum or proximal jejunum [6,7], and released from there into the blood. However, the supply of zinc by our diet is dependent on its amount and bioavailability. It has been estimated that in a western mixed diet, this bioavailability is about 20–30% of total contained zinc [8]. Various agents can decrease zinc absorption [9]; among them are phytates [10,11] and other metals such as copper and, to a lesser extent, calcium and iron [12–15]. Based on average bioavailability, the recommended daily intakes of zinc range from 10 to 15 mg in adults but may be higher under certain circumstances, such as pregnancy and during lactation, where an extra 5–10 mg may be required.

Within the body, two pools of zinc were identified. The majority of zinc (about 90%) is relatively slowly exchanged with the blood plasma and, for example, concentrated in the bones. The remaining 10% of zinc however is rapidly exchanged and it is this pool of zinc that needs daily replenishment and that is therefore especially reactive to the amount of zinc absorbed in the GI system. Exchange of zinc across membranes is mediated by two solute-linked carrier (SLC) families, the SLC30A (ZnT-1 to ZnT-10) and the SLC39A (Zrt, Irt-like protein ZIP1 to ZIP14) family. These transporters show tissue-specific expression and localize to distinct subcellular compartments, where, in general, ZnT proteins transport zinc out of the cytosol and Zip proteins move zinc into the cytosol.

Unfortunately, the clinical diagnosis of a zinc deficiency in humans currently faces major limitations [16]. Although measuring zinc concentrations in blood plasma or serum is currently the most commonly used method, this provides only a snapshot of the zinc status of an individual, and given that serum zinc concentrations may fluctuate by as much as 20% during a day [17], a single blood drawing has low validity. Alternatively, assessment of zinc levels in hair or nail samples might be a preferable method as an average of zinc levels over a longer period of hair or nail growth will be evaluated. The lack of generally accepted biomarkers of zinc status complicates the assessment of zinc deficiency. Thus, zinc deficiency in humans is probably commonly overlooked, especially if only mild or transient and the exact prevalence is currently unknown [18]. In addition, the symptoms of mild zinc deficiency are much less dramatic and more diffuse than those observed in AE. Mild zinc deficiency, for example, might

http://dx.doi.org/10.5772/63203

not cause typical skin lesions. Nevertheless, current estimates are that about 17.3% of the global population is at risk of developing zinc deficiency [19].

Given this relatively high number of potentially affected individuals, in the following, on the background of nutritional deficiencies, we will provide a more detailed discussion of the role of zinc in the body and its association with specific pathologies.

2. The role of zinc in the body

2.1. Zinc and the endocrine system

The endocrine system is comprised of a number of glands in the body and includes the ovaries, the testes, and the thyroid, parathyroid, adrenal, and pituitary glands. Further, the pineal body, the pancreas, and specific cells releasing hormones in the GI tract, kidney, heart, and placenta are part of this system. Zinc has manifold influences on the endocrine system (**Figure 1**). Among them, a role in the metabolism of androgen hormones, estrogen, and progesterone, together with the prostaglandins, a role in the secretion of insulin, and a role in the regulation of thymic hormones have been reported.

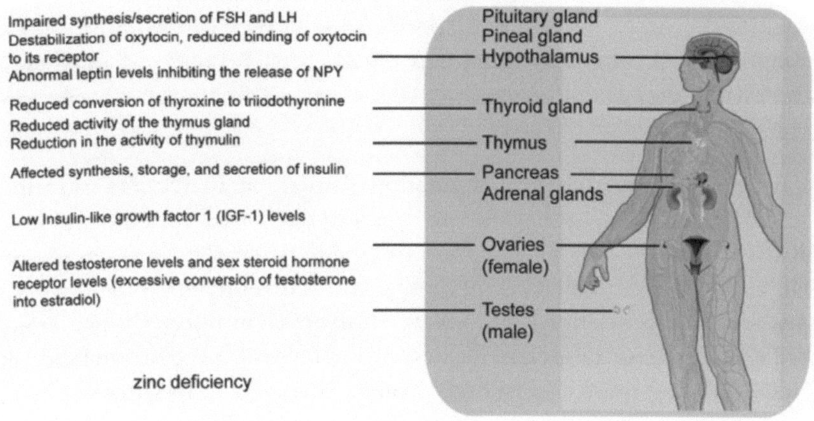

Figure 1. Overview of the major effects of zinc deficiency on the endocrine system. FSH, follicle-stimulating hormone; LH, luteinizing hormone; NPY, neuropeptide Y.

An involvement of zinc in the regulation of sex hormones in males and females can be concluded indirectly, as zinc deficiency in pregnancy is associated with disruption of the estrous cycle, frequent spontaneous abortion, extended pregnancy or prematurity, inefficient labor, and atonic bleeding [20,21]. On a molecular level, zinc deficiency in the female can lead to impaired synthesis/secretion of follicle-stimulating hormone (FSH) and luteinizing hormone (LH). Moreover, the nuclear receptors for sex steroids contain zinc finger motifs in their protein structure that might explain zinc dependency of these hormonal systems. In addition, zinc deficiency alters testosterone levels and modifies sex steroid hormone receptor levels [22].

Zinc deficiency affects the enzyme complex aromatase that is normally inhibited by zinc. The increased activation results in an excessive conversion of testosterone into estradiol. Further, zinc deficiency and the depletion of testosterone result in an inhibition of spermatogenesis [23] in males.

Hypogonadism is a major manifestation of zinc deficiency. Although steroids might play a role, it was speculated that the hypogonadism is due to a hypothalamic dysfunction and associated with low insulin-like growth factor 1 (IGF-1) levels. Testicular development can be rescued by zinc supplementation [24].

Zinc also improves the stability of oxytocin, and the stabilization effect is correlated with the ability of the divalent metal to interact with oxytocin. Zinc is essential for the binding of oxytocin to its cellular receptor [25,26].

Zinc plays a key role in the synthesis, storage, and secretion of insulin. Zinc is able to stimulate the action of insulin and insulin receptor tyrosine kinase. A low zinc status has been associated with diabetes (types 1 and 2) [27] and zinc supplementation was reported capable of restoring insulin secretion [28].

In mammals, insulin is synthesized in islets of Langerhans, made of four different cell types in the pancreas. The majority are insulin-producing β cells. There, insulin is stored in a zinc-containing hexameric form. Zinc deficiency can affect the ability of β cells to produce and secrete insulin [29]. ZnT-8 (*SLC30A8*), a specific zinc transporter found localized at insulin secretory granules in β cells, was identified [30,31]. ZnT-8 knockout mice show impaired insulin secretion [32], and *SLC30A8* variants, reducing ZnT-8 activity, increase type 2 diabetes risk in humans [33].

Zinc also plays an important role in the regulation of nutrition. In humans and animals models, marginal zinc deficiency has been shown to result in decreased appetite and low body mass [34], features that can not only be observed in zinc deficiency but also in anorexia nervosa patients [35]. Although the underlying mechanisms are currently not well understood, changes in neurotransmitter concentrations and synaptic transmission at the hypothalamic level might be associated with decreased appetite. Recent studies have also shown a relationship between zinc and leptin [36]. Leptin is a hormonal protein, involved in features such as satiety and energy balance, but was also reported as anti-obesity factor. The major target organ for leptin is the hypothalamus, where leptin controls food intake through its receptors, inhibiting the release of neuropeptide Y (NPY) which has an augmentative effect on food intake [37]. However, leptin also plays a role in immunity [38].

Further, a correlation between zinc deficiency in geriatric patients and reduced activity of the thymus gland and thymic hormones has been reported [39]. Zinc promotes the conversion of thyroid hormones thyroxine to triiodothyronine and zinc deficiency can result in hypothyroidism, a common disorder of the endocrine system characterized by decreased production of thyroid hormone.

Finally, in animal models, zinc deficiency leads to a reduction in the activity of thymulin. Thymulin is a nonapeptide produced by thymic epithelial cells that requires zinc for its biological activity [40]. Zinc is bound to thymulin in a 1:1 stoichiometry [41], which results in a conformational change generating the active form of thymulin [42]. Thymulin is required for maturation of T-helper cells, leading to a decrease in T-helper 1 (Th(1)) cytokines in zinc deficiency [40]. This last example for a role of zinc in the endocrine system already hints toward its additional involvement in regulatory processes in the immune system. There, zinc ions play a role by binding to specific proteins but also as a second messenger regulating signal transduction in various kinds of immune cells [39].

2.2. Zinc and the immune system

The zinc status of an individual affects the majority of immunological events such as hematopoiesis, immune cell function and survival, humoral immunity, and cytokine secretion [43–45]. This results, for example, in an increased susceptibility to infections with a decreasing zinc status as reported from animal models and human studies [46]. Similarly, in AE, one hallmark symptom is a reduced immune function [47] that is visible through the atrophy of the thymus, functional impairment and reduced numbers of lymphocytes, and increased susceptibility to infections [48]. In contrast, beneficial effects of zinc supplementation have been found regarding the incidence and duration of acute and persistent diarrhea [49–51], incidence of acute lower respiratory infections [52], and duration of the common cold [53].

The immune response can be divided into two major mechanisms: innate and adaptive immunity [54]. Zinc is involved in virtually all aspects of innate and adaptive immunity and it is therefore not surprising that it has been reported that zinc deficiency results in a depressed immune system (**Figure 2**). The effects of zinc on both mechanisms are based on the myriad roles for zinc in basic cellular functions such as DNA replication, RNA transcription, proliferation and differentiation, and immune cell activation.

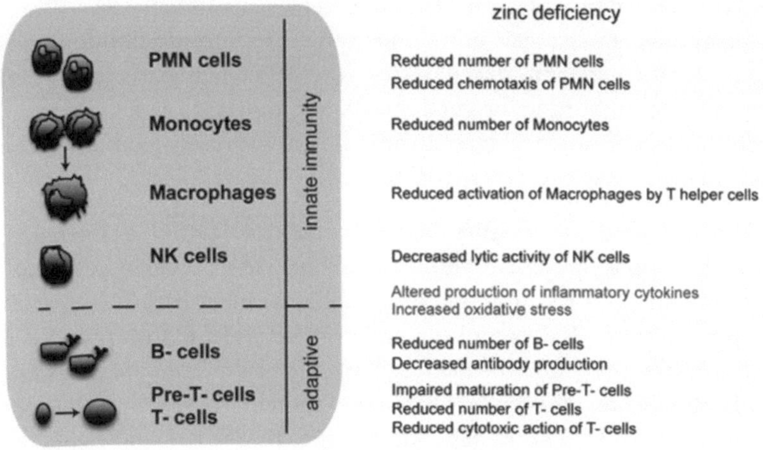

Figure 2. Overview of the major effects of zinc deficiency on the immune system. PMN, polymorphonuclear; NK, natural killer.

Under physiological conditions, the normal response of the body against pathogens is initiated by the activation of the complement system, macrophages, natural killer (NK) cells, and polymorphonuclear (PMN) cells. Under zinc-deficient conditions, all these defending mechanisms are affected [55–58]. For example, PMN are the first cells to actively enter an infected tissue and their chemotaxis is reduced under zinc deficiency [59]. Moreover lytic activity of NK cells is decreased during zinc deficiency.

During the acute phase, profound changes in zinc homeostasis occur, which may serve multiple purposes. For example, altered zinc levels may serve as signaling factor, but zinc redistribution that leads to a decrease in plasma zinc levels might also be used actively by the immune system to attack certain intracellular pathogens. In addition reduced zinc levels lead to a shift of leukopoiesis toward the generation of myeloid cells [44] and reduced extracellular zinc may increase monocyte differentiation [60].

Adaptive immunity is mediated by T and B cells. Zinc deficiency was reported to alter the number and function of neutrophil granulocytes, monocytes, and NK, T, and B cells [39]. In particular, the vulnerability to infections is associated with an impaired T- and B-lymphocyte development and activity [61–63]. T-cell progenitors mature in the thymus and zinc deficiency causes thymic atrophy (see Section 2.1). During maturation, pre-T cells have been reported to be most susceptible to zinc deficiency, which resulted in a loss of up to 50% of these cells in a mouse model [44]. Mature T-helper cells promote the functions of other immune cells. For example, T-helper cells play a role in the activation of macrophages. Cytotoxic T cells, in turn, have a more direct role in the immune response by eliminating virus-infected cells and tumor cells.

In addition, zinc deficiency leads to a reduction of premature and immature B cells and decreases antibody production [64].

Further, zinc plays a role in the inflammatory response and in particular in the termination of this process. Inflammatory processes, in particular sepsis, are associated with major changes in zinc metabolism and homeostasis and accompanied by a redistribution of zinc between tissues. This is underlined by data showing that zinc deficiency leads to aggravated inflammation, increased bacterial burden, organ damage, and mortality in a mouse model of sepsis [65,66]. Similarly, in human patients belonging to a group with elevated risk for zinc deficiency, a high incidence of sepsis has been reported [67].

Zinc is able to downregulate the production of inflammatory cytokines mediated via the NF-kB (nuclear factor "kappa-light-chain enhancer" of activated B cells) pathway [68]. NF-kB mediates the expression of pro-inflammatory cytokines, including TNF-α and IL-1b. Zinc inhibits the I kappa B kinase (IKK) complex member IKKβ [69] and thus prevents downstream translocation of NF-kB dimers into the nucleus [70], where they can increase gene expression by binding to kB sites located within the promoter region of, for example, interleukin 6 (IL-6) [71]. In monocytes, NF-kB activation depends on zinc. Under zinc-deficient conditions, this anti-inflammatory feature is absent possibly contributing to prolonged or chronic inflammation. However, there might be cell-type-specific responses of NF-kB signaling to zinc [39]. Moreover, immune cells contain a vast number of different zinc-dependent enzymes / proteins

http://dx.doi.org/10.5772/63203

that, so far, have not been studied in detail but are likely involved in the immunomodulatory functions of zinc.

There is a close relation between inflammatory processes and oxidative stress [43]. Under physiological conditions, zinc itself is not an antioxidant, because it does not participate in redox reactions, but is considered a "proantioxidant" [72] since it protects cells from the damaging effect of oxygen radicals generated during immune activation [73]. For example, zinc release from thiolate bonds can prevent lipid peroxidation [74] and release from metallothioneins (MTs), major zinc-buffering proteins in cells, may reduce membrane damage by free radicals during inflammation. Zinc deficiency causes an elevation of oxidative stress and zinc supplementation was shown to decrease markers of oxidative stress [43].

Intriguingly, changes in immune function similar to those seen during zinc deficiency were observed in immunosenescence [75]. For example, thymic atrophy, increased inflammation, impaired cellular and humoral immune responses, and recurrent infections were observed [76]. These observations suggest that age-dependent zinc deficiency could be a contributing factor to the age-related decline in immune system function.

2.3. Zinc and the brain

Zinc is one of the most abundant trace metals in the brain. There, different pools of zinc can be found. More than 80% of brain zinc is bound to proteins regulating their protein structure, participating in signaling, or acting as cofactor of enzymes. Free (aqueous) zinc exists predominantly within synaptic vesicles of presynaptic terminals of glutamatergic (zincergic) neurons mostly in the hippocampus, amygdala, and cerebral cortex serving as signaling ion and neuromodulator. During development, brain zinc concentration constantly rises until adulthood where zinc levels remain constant at around 200 μM total brain zinc [77]. During neonatal development the highest zinc levels can be found in the cerebellum, which is accompanied by the rapid growth due to the necessity for motor skills acquisition during this developmental stage. In the adult brain, zinc is especially enriched in the hippocampus and cerebral cortex. On cellular level, in neurons, zinc is diffusely distributed in the cytoplasm and nucleus [78] mainly bound to proteins. However, zinc is also found in neuronal processes [78] and in vesicles of presynaptic terminals [79,80].

The brain zinc homeostasis is tightly controlled by transporters at the blood-brain barrier (BBB) and by intracellular regulatory system. In order to cross the BBB, an active transport of zinc is required. So far the mechanism behind zinc uptake is not fully understood but zinc transport might be mediated by L-histidine [81] and the membrane transporter DMT1 (divalent metal transporter 1) [82]. Further, the export of zinc into brain extracellular fluid remains to be deciphered. In brain cells, zinc homeostasis is controlled by zinc transporters and small zinc-binding proteins. Proteins of the ZnT family transport zinc out of the cytoplasm into organelles or out of the cell, and ZIPs transport zinc into the cytoplasm. Further, intracellular zinc homeostasis is regulated by small zinc-binding proteins like the MTs. Among them MT3 is brain specific.

Brain development is a tightly controlled and highly concerted succession of processes including proliferation, differentiation, apoptosis, maturation, migration, myelination as well as synaptogenesis, and pruning. Animal experiments have shown that zinc is involved in all these processes and that zinc deficiency during early embryonal development has teratogenic effects and affects all organ system but the brain seems to be particularly vulnerable. It is noteworthy that consequences of zinc deficiency strongly depend on the degree of severity as well as the time of onset during development. For example, severe zinc deficiency during early developmental stages is associated with neural tube closure deficits [83,84] and brain malformations. Marginal zinc deficiency during the whole course of pregnancy in turn affects the proliferation of neural progenitor cells, the expression of N-methyl-D-aspartate receptor (NMDAR), and growth and transcription factors which in turn affect the regulation of signaling pathways involved in brain development and function in the offspring of zinc-deficient mothers [85–87]. If zinc deficiency occurs during postnatal development, pups showed a decreased brain size; reduced brain DNA, RNA, and protein concentrations [88] as well as decreased numbers of neurons and impaired dendritic arborization in the cerebellum of Purkinje, basket, stellate, and granule cells [89–91]; and decreased numbers of progenitor cells in the dentate gyrus.

The number of stem cells is reduced in the offspring of zinc-deficient animals [92]. Further, in vivo and in vitro studies have shown that zinc deficiency leads to a decrease in progenitor cell proliferation, most likely due to the arrest of the cell cycle whereby cell proliferation is inhibited. In addition, the amount of spontaneous apoptosis is increased under zinc-deficient conditions [87,93,94]. Similarly, knockdown of the zinc transporter Zip12 leads to disturbances in neuronal differentiation affecting neurite sprouting and outgrowth, neurulation, and embryonic development [84]. These features were accompanied by a reduced tubulin polymerization that was observed in Zip12 knockdown mice as well as in pups of zinc-deficient mothers [84,95,96]. Given that transcription factors like nuclear factor of activated T-cells and NF-kB need a functional cytoskeleton for nuclear shuttling [86,97], it is not surprising that prenatal zinc deficiency leads to a deregulation of transcription factors in neurons that are crucial for brain development through regulating the expression of genes that are involved in proliferation, differentiation, and synaptic plasticity [86].

Even if the zinc deficiency occurs only during pre- or early postnatal development and zinc levels are completely restored at adulthood, long-term effects can be observed in the offspring of zinc-deficient mothers. Prenatal zinc-deficient mice showed impaired maternal behavior, impaired auditory discrimination, alterations in ultrasonic vocalizations of adult males to female urine, increased aggressiveness and emotionality [98–101], severe learning deficits and memory impairments [102–105], and enhanced stress response [106,107] when they were tested as adults. Given the fact that these behavioral alterations observed in prenatal zinc-deficient animals resemble behavioral patterns of people suffering from neurodevelopmental and neuropsychiatric disorders like autism spectrum disorders, schizophrenia, and depression, a role of zinc deficiency in the etiology of these disorders is possible (see below).

Zinc is not only needed during brain development but also to maintain proper brain function. As already mentioned, free zinc is stored in presynaptic vesicles together with glutamate in zincergic neurons of the hippocampus, amygdala, and cerebral cortex. During synaptic activity, zinc is released and serves as signal ion. Enriched in the synaptic cleft, zinc can modulate synaptic signal transduction via the modulation of glutamate receptors, ion channels, cell adhesion molecules, and the pre- or postsynaptic uptake of calcium [82] or may directly act as neurotransmitter via metabotropic GPR39 receptor (GPR39) [108] that is involved in glutamatergic transmission [109]. For example, zinc can allosterically inhibit NMDA receptors through binding to a subunit at low levels or inhibit NMDAR in a voltage-dependent manner at high levels [110] and therefore modulate the signal transduction at the postsynaptic site. Additionally, zinc is also able to modulate α-amino-3-hydroxy-5-methyl-4-isoxazolepropionic acid receptor properties [111,112]. Through the mentioned receptors and voltage-dependent calcium channels, zinc can also enter the postsynaptic neuron and increase the intracellular zinc concentration, which might serve as additional signal. For example, the availability of zinc within the postsynaptic neuron is necessary for, and modulates the assembly of, the postsynaptic scaffold where receptors are anchored [98,113]. Furthermore, the increase of intracellular zinc concentration affects and sometimes is necessary for the induction of long-term potentiation (LTP) through the modulation of kinases and neurotrophic factors activity [114,115]. LTP is believed to be the molecular process underlying memory formation indicating that the increase of cytosolic zinc plays an important role in learning and cognitive performance [116]. However, zinc is not only taken up by the postsynaptic neuron but also by the presynaptic neuron where it might act in a negative feedback mechanism preventing further glutamate release [117,118].

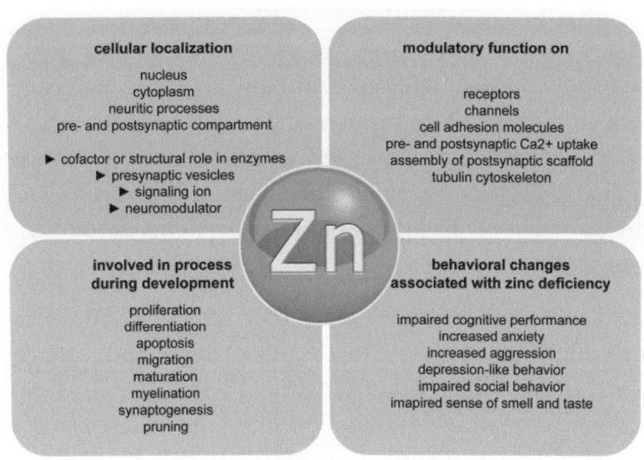

Figure 3. Overview of the major functions of zinc in the brain and effects of zinc deficiency.

Given the multifaceted action of zinc in the brain (**Figure 3**), it is not surprising that zinc deficiency leads to alterations in cognitive performance and behavior in animal models and possibly humans. Acute zinc-deficient animals demonstrated impaired learning and memory behavior [119–122], increased aggressiveness [123] and hyperreactivity/irritability [98], and anxiety as well as depressive-like behavior [109,124]. Additionally, neurosensory functions like smell and taste are impaired through zinc deficiency [125].

The dysregulation of zinc homeostasis is a common and well-investigated feature in neurodegenerative disorders like Alzheimer's disease (AD) or Parkinson's disease (PD), but only little is known about the role of zinc in the normal aging process of the brain although elderly people frequently suffer from zinc deficiency [126,127] that might be due to lower dietary intake, chronic inflammation, and an age-related decline in zinc transport mechanisms [128,129]. A decrease in histochemically reactive zinc has also been reported in aged animals [130]. Additionally, in a rodent model of aging, zinc was less distributed in the hippocampus, and the expression of ZnT3, a zinc transporter responsible for the transport of cytosolic zinc into synaptic vesicles, was significantly reduced [131]. Given that zinc is important in synaptic plasticity, the underlying cellular mechanism of learning and memory and a lower availability of zinc in the aging brain might affect this feature.

Taken together, zinc has a pivotal role in the brain during all stages of life. Zinc deficiency during brain development leads to persistent deficits in cognitive functions and behavior. In adults, zinc deficiency results in disturbed behavior and may contribute to the age-dependent decline of cognitive performance. Therefore, alterations in zinc homeostasis are intensively investigated in brain disorders but zinc deficiency affects all organ systems of the body and therefore can lead to a plethora of disorders.

3. Zinc deficiency and associated disorders

Zinc deficiency itself is associated with several clinical signs. For example, marginal zinc deficiency may result in depressed immunity, impairment of memory, neurosensory problems such as impaired taste and smell as well as onset of night blindness, and decreased spermatogenesis in males [46,132]. Severe zinc deficiency is characterized by a more severely depressed immune function resulting in frequent infections, dermatitis, diarrhea, alopecia, and mental disturbances [46].

In 1961 it was hypothesized for the first time that human nutritional zinc deficiency of environmental origin might be associated with a disorder in the form of adolescent dwarfism. Patients found in Iranian villages that consumed a severely restricted (inadequate) diet consisting mainly of wheat bread with animal protein food sources largely absent displayed growth retardation and hypogonadism and were iron deficient [133]. Dwarfism and absent sexual maturation were assumed to be caused by zinc deficiency. Likewise in 1967 zinc deficiency was reported as the etiological factor responsible for retarded sexual development and growth in adolescents from rural Egypt with similar dietary habits [1,2,134]. Zinc supplementation resulted in the subject's growth and in the development of their genitalia.

In these cases, zinc deficiency was based on nutrition. It is known from congenital defects like mutations in the zinc transporter ZIP4 that severe zinc deficiency can even become lethal.

However, zinc deficiency is also associated with specific disorders. Few of them will be briefly discussed here exemplarily. Alterations in zinc homeostasis have been implicated in various neurodegenerative, neurological, and neuropsychological disorders such as mood disorders, autism spectrum disorders (ASD), AD, PD, Huntington's disease (HD), multiple sclerosis (MS), and amyotrophic lateral sclerosis (ALS) [135,136].

Probably the strongest association of acute zinc deficiency and a brain disorder is found regarding depression. Therefore, this topic is discussed in a separate chapter in this book.

Zinc deficiency during brain development in turn seems to result in a different clinical picture, arguing for multiple roles of zinc in brain development and adult brain function. For example ASD, a group of neurodevelopmental disorders characterized by the core features of impairments in social interaction and communication as well as repetitive and restrictive behavior [137] might develop influenced by both a genetic component and non-genetic factors such as zinc deficiency during development and early in life [138,139]. Accordingly, investigation of zinc levels in autistic children using hair samples revealed a high prevalence rate for zinc deficiency especially in the youngest age group (0–3 years) [140,141]. Rodent animal models of embryonic zinc deficiency display depending on the severity of zinc deficiency autism-like behavior as well as behavioral symptoms associated with common comorbidities of ASD [98]. Besides memory and learning deficits as well as impaired social behavior in prenatal and perinatal zinc-deficient mice and rats [142], in mouse models of acute and prenatal zinc deficiency, hyper-responsiveness, seizures, and impaired ultrasonic vocalization were observed.

However, besides zinc deficiency causing or contributing to the etiology of a disorder, zinc deficiency that can be systemic or local can also be the consequence. In AD, for example, the most common form of dementia that is characterized by extracellular amyloid plaques consisting of amyloid-β (Aβ) polymers and intracellular neurofibrillary tangles composed of tau protein, zinc is sequestered by the zinc-binding Aβ peptides into extracellular senile plaques. Due to the importance of zinc for proper brain functionality and high prevalence of zinc deficiency among elderly people, the potential role of zinc as a contributing and modifying factor in the course of AD moved into the spotlight of research. While there has been emerging evidence of abnormalities in AD regarding brain zinc, copper, and iron homeostasis, there seems to be a lack of consensus regarding alterations in peripheral zinc status [143]. While several studies reported a significant decrease in serum zinc level of AD patients, others detected a significant increase while some found no difference between patients and controls [144]. One of the possible roles of zinc in AD is its involvement in Aβ accumulation. In plaques both copper and zinc are able to bind Aβ directly. Thus, analysis of Aβ plaques of postmortem AD brain reveals high concentrations of accumulated zinc and copper [143]. Since Aβ is proteolytically cleaved from the amyloid precursor protein (APP), a possible role of zinc in APP synthesis and processing was investigated. APP is cleaved in its Aβ region by α-secretase leading to the production of soluble amyloid precursor peptide (sAA) [145] before by cleavage with β- and γ-secretase, the Aβ peptide is formed [144,146]. In APP's ectodomain, which includes the position of the cleavage site of α-secretase, the enzyme responsible for the first processing step, a zinc-binding site is localized [147] suggesting a possible influence of zinc on the secretase's activity [144]. In case of zinc deficiency, not only APP cleavage but also synaptic transmission and plasticity might be impaired by sequestration of zinc [148,149].

Besides the brain, high concentration of zinc compared to other tissues can be found in pancreatic tissue, suggesting a role in endocrine and exocrine function of the organ [150,151]. Most importantly, zinc plays a role in synthesis, secretion, and signaling of insulin [152]. Due to its various functions in the pancreas, alterations of zinc homeostasis have been implicated in the pathogenesis of diabetes and in impaired insulin sensitivity [151,153]. Furthermore, hyperzincuria and hypozincemia are frequently diagnosed in diabetic patients [151,153]. Assessment of serum zinc levels in a set of type 1 and type 2 diabetic patients in comparison to healthy controls revealed significant lower mean serum zinc levels in the diabetic groups [154]. db/db mice, an animal model for obesity and diabetes, exhibit hyperglycemia, hyperinsulinemia, hyperleptinemia, and obesity. After dietary zinc supplementation they showed an attenuated fasting hyperinsulinemia and hyperglycemia and elevated pancreatic zinc levels. In db/db mice on a zinc-deficient diet, an opposite effect was observed indicating a possible connection between glycemic control and zinc [155]. Likewise in human studies, beneficial effects of zinc in diabetes type 1 and type 2 have been described [156,157].

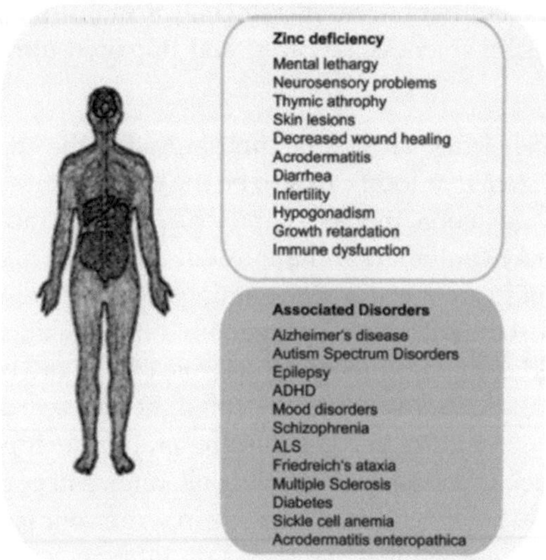

Figure 4. Symptoms of zinc deficiency and associated disorders. ADHD, attention deficit hyperactivity disorder; ALS, amyotrophic lateral sclerosis. Intriguingly, a high number of brain disorders seem to be associated with zinc deficiency.

Taken together, due to the manifold roles of zinc in various organ systems as a key structural component of enzymes and proteins but also signaling ion, zinc deficiency, depending on the time of onset, duration, severity, and systemic or tissue-specific nature, can result in a variety of symptoms with different severities. Further, zinc deficiency is associated with several disorders, probably acting as trigger, modifier, or even cause (**Figure 4**). However, the

http://dx.doi.org/10.5772/63203

molecular mechanisms how zinc deficiency contributes to specific phenotypes are currently not well understood.

4. Conclusions

Although only recently recognized, the importance of zinc as essential trace metal in the body and in particular as signaling molecule is substantiated greatly. Zinc is a trace element with various roles in physiological processes (Kaur et al., 2014). It has been used as a drug in the prevention and treatment of some diseases and new strategies for more targeted delivery or modification of zinc signaling are promising future therapeutic approaches, especially in brain disorders (Ayton et al., 2015; Lee et al., 2015). In addition, the present knowledge about zinc signaling in the various processes and involved pathways seems to be disconnected by specific types of zinc signal used, with different kinetics and sources of zinc. However, most likely, an interplay between the different systems described above may exist by common underlying principles of zinc signaling.

The high rate of the world's population that is at risk for inadequate zinc supply underlines the need for further research on zinc signaling and the need for public health programs to control zinc deficiency.

The authors were supported by the Else Kröner-Fresenius Stiftung (AMG), Juniorprofessuren program of the state of Baden-Württemberg (AMG, AKS), and Evangelisches Studienwerk e.V. Villigst (SH).

Author details

Ann Katrin Sauer, Simone Hagmeyer and Andreas M. Grabrucker*

*Address all correspondence to: andreas.grabrucker@uni-ulm.de

WG Molecular Analysis of Synaptopathies, Neurology Department, Neurocenter of Ulm University, Ulm, Germany

References

[1] Prasad AS, Miale A, Farid Z, Sandstead HH, Schulert AR: Zinc metabolism in patients with the syndrome of iron deficiency anemia, hepatosplenomegaly, dwarfism, and hypognadism. J Lab Clin Med 1963;61:537–549.

[2] Prasad AS, Schulert AR, Miale A, Farid Z, Sandstead HH: Zinc and iron deficiencies in male subjects with dwarfism and hypogonadism but without ancylostomiasis, schistosomiasis or severe anemia. Am J Clin Nutr 1963;12:437–444.

[3] Küry S, Kharfi M, Blouin E, Schmitt S, Bézieau S. Clinical utility gene card for: acrodermatitis enteropathica – update 2015. Eur J Hum Genet 2016;24(5).

[4] Küry S, Dréno B, Bézieau S, Giraudet S, Kharfi M, Kamoun R, Moisan JP: Identification of SLC39A4, a gene involved in acrodermatitis enteropathica. Nat Genet 2002;31(3): 239–240.

[5] Kasana S, Din J, Maret W: Genetic causes and gene–nutrient interactions in mammalian zinc deficiencies: acrodermatitis enteropathica and transient neonatal zinc deficiency as examples. J Trace Elem Med Biol 2015;29:47–62.

[6] Krebs NF, Westcott JE, Huffer JW, Miller LV: Absorption of exogenous zinc and secretion of endogenous zinc in the human small intestine. FASEB J 1998;12:A345.

[7] Lee HH, Prasad AS, Brewer GJ, Owyang C: Zinc absorption in human small intestine. Am J Physiol 1989;256(1 Pt 1):G87–G91.

[8] Gallaher DD, Johnson PE, Hunt JR, Lykken GI, Marchello MJ: Bioavailability in humans of zinc from beef: intrinsic vs extrinsic labels. Am J Clin Nutr 1988;48(2):350–354.

[9] Cousins RJ: Gastrointestinal factors influencing zinc absorption and homeostasis. Int J Vitam Nutr Res 2010;80(4–5):243–248.

[10] Lönnerdal B: Dietary factors influencing zinc absorption. J Nutr 2000;130(5S Suppl): 1378–1383.

[11] King JC: Determinants of maternal zinc status during pregnancy. Am J Clin Nutr 2000;71(5 Suppl):1334–1343.

[12] Wood RJ, Zheng JJ: High dietary calcium intakes reduce zinc absorption and balance in humans. Am J Clin Nutr 1997;65(6):1803–1809.

[13] Argiratos V, Samman S: The effect of calcium carbonate and calcium citrate on the absorption of zinc in healthy female subjects. Eur J Clin Nutr 1994;48(3):198–204.

[14] O'Brien KO, Zavaleta N, Caulfield LE, Wen J, Abrams SA: Prenatal iron supplements impair zinc absorption in pregnant Peruvian women. J Nutr 2000;130(9):2251–2255.

[15] Fischer Walker C, Kordas K, Stoltzfus RJ, Black RE: Interactive effects of iron and zinc on biochemical and functional outcomes in supplementation trials. Am J Clin Nutr 2005;82(1):5–12.

[16] Roohani N, Hurrell R, Kelishadi R, Schulin R: Zinc and its importance for human health: an integrative review. J Res Med Sci 2013;18(2):144–157.

[17] Hambidge KM, Goodall MJ, Stall C, Pritts J: Post-prandial and daily changes in plasma zinc. J Trace Elem Electrolytes Health Dis 1989;3(1):55–57.

[18] Willoughby JL, Bowen CN: Zinc deficiency and toxicity in pediatric practice. Curr Opin Pediatr 2014;26(5):579–584.

[19] Wessells KR, Brown KH: Estimating the global prevalence of zinc deficiency: results based on zinc availability in national food supplies and the prevalence of stunting. PLoS One 2012;7(11):50568.

[20] Jameson S: Zinc status in pregnancy: the effect of zinc therapy on perinatal mortality, prematurity, and placental ablation. Ann N Y Acad Sci 1993;678:178–192.

[21] Favier AE: The role of zinc in reproduction. Hormonal mechanisms. Biol Trace Elem Res 1992;32:363–382.

[22] Om AS, Chung KW: Dietary zinc deficiency alters 5 alpha-reduction and aromatization of testosterone and androgen and estrogen receptors in rat liver. J Nutr 1996;126(4): 842–848.

[23] Bedwal RS, Bahuguna A: Zinc, copper and selenium in reproduction. Experientia 1994;50(7):626–640.

[24] Karaca Z, Tanriverdi F, Kurtoglu S, Tokalioglu S, Unluhizarci K, Kelestimur F: Pubertal arrest due to Zn deficiency: the effect of zinc supplementation. Hormones (Athens) 2007;6(1):71–74.

[25] Liu D, Seuthe AB, Ehrler OT, Zhang X, Wyttenbach T, Hsu JF, Bowers MT: Oxytocin-receptor binding: why divalent metals are essential. J Am Chem Soc 2005;127(7):2024–2025.

[26] Avanti C, Hinrichs WL, Casini A, Eissens AC, Dam AV, Kedrov A, Driessen AJ, Frijlink HW, Permentier HP: The formation of oxytocin dimers is suppressed by the zinc–aspartate–oxytocin complex. J Pharm Sci 2013;102(6):1734–1741.

[27] DiSilvestro RA: Zinc in relation to diabetes and oxidative disease. J Nutr 2000;130(5S Suppl):1509–1511.

[28] Cunningham JJ: Micronutrients as nutriceutical interventions in diabetes mellitus. J Am Coll Nutr 1998;17(1):7–10.

[29] Chausmer AB: Zinc, insulin and diabetes. J Am Coll Nutr 1998;17(2):109–115.

[30] Chimienti F, Devergnas S, Favier A, Seve M: Identification and cloning of a beta-cell-specific zinc transporter, ZnT-8, localized into insulin secretory granules. Diabetes 2004;53(9):2330–2337.

[31] Rutter GA, Chabosseau P, Bellomo EA, Maret W, Mitchell RK, Hodson DJ, Solomou A, Hu M. Intracellular zinc in insulin secretion and action: a determinant of diabetes risk? Proc Nutr Soc 2016;75(1):61–72.

[32] Pound LD, Sarkar SA, Benninger RK, Wang Y, Suwanichkul A, Shadoan MK, Printz RL, Oeser JK, Lee CE, Piston DW, McGuinness OP, Hutton JC, Powell DR, O'Brien RM:

Deletion of the mouse Slc30a8 gene encoding zinc transporter-8 results in impaired insulin secretion. Biochem J 2009;421(3):371–376.

[33] Solomou A, Meur G, Bellomo E, Hodson DJ, Tomas A, Li SM, Philippe E, Herrera PL, Magnan C, Rutter GA: The zinc transporter Slc30a8/ZnT8 is required in a subpopulation of pancreatic α-cells for hypoglycemia-induced glucagon secretion. J Biol Chem 2015;290(35):21432–21442.

[34] El-Shazly AN, Ibrahim SA, El-Mashad GM, Sabry JH, Sherbini NS: Effect of zinc supplementation on body mass index and serum levels of zinc and leptin in pediatric hemodialysis patients. Int J Nephrol Renovasc Dis 2015;8:159–163.

[35] Tannhauser PP: Anorexia nervosa: a multifactorial disease of nutritional origin? Int J Adolesc Med Health 2002;14(3):185–191.

[36] Baltaci AK, Mogulkoc R, Halifeoglu I: Effects of zinc deficiency and supplementation on plasma leptin levels in rats. Biol Trace Elem Res 2005;104:41–46

[37] Levenson CW: Zinc regulation of food intake: new insights on the role of neuropeptide Y. Nutr Rev 2003;61(7):247–249.

[38] Baltaci AK, Mogulkoc R: Leptin and zinc relation: in regulation of food intake and immunity. Indian J Endocrinol Metab 2012;16(Suppl 3):611–616.

[39] Haase H, Rink L: Zinc signals and immune function. Biofactors 2014;40(1):27–40.

[40] Prasad AS: Zinc: mechanisms of host defense. J Nutr 2007;137(5):1345–1349.

[41] Cung MT, Marraud M, Lefrancier P, Dardenne M, Bach JF, Laussac JP: NMR study of a lymphocyte differentiating thymic factor. An investigation of the Zn(II)-nonapeptide complexes (thymulin). J Biol Chem 1988;263(12):5574–5580.

[42] Dardenne M, Pléau JM, Nabarra B, Lefrancier P, Derrien M, Choay J, Bach JF: Contribution of zinc and other metals to the biological activity of the serum thymic factor. Proc Natl Acad Sci U S A 1982;79(17):5370–5373.

[43] Prasad AS: Zinc: role in immunity, oxidative stress and chronic inflammation. Curr Opin Clin Nutr Metab Care 2009;12(6):646–652.

[44] Fraker PJ, King LE: Reprogramming of the immune system during zinc deficiency. Annu Rev Nutr 2004;24:277–298.

[45] Foster M, Samman S: Zinc and regulation of inflammatory cytokines: implications for cardiometabolic disease. Nutrients 2012;4(7):676–694.

[46] Shankar AH, Prasad AS: Zinc and immune function: the biological basis of altered resistance to infection. Am J Clin Nutr 1998;68(2 Suppl):447–463.

[47] Moynahan EJ: Letter: zinc deficiency and cellular immune deficiency in acrodermatitis enteropathica in man and zinc deficiency with thymic hypoplasia in fresian calves: a possible genetic link. Lancet 1975;2(7937):710.

[48] Wellinghausen N, Rink L: The significance of zinc for leukocyte biology. J Leukoc Biol 1998;64(5):571–577.

[49] Sazawal S, Black RE, Bhan MK, Jalla S, Bhandari N, Sinha A, Majumdar S: Zinc supplementation reduces the incidence of persistent diarrhea and dysentery among low socioeconomic children in India. J Nutr 1996;126(2):443–450.

[50] Sazawal S, Black RE, Bhan MK, Jalla S, Sinha A, Bhandari N: Efficacy of zinc supplementation in reducing the incidence and prevalence of acute diarrhea – a community-based, double-blind, controlled trial. Am J Clin Nutr 1997;66(2):413–418.

[51] Walker CL, Black RE: Zinc for the treatment of diarrhoea: effect on diarrhoea morbidity, mortality and incidence of future episodes. Int J Epidemiol 2010;39(1):i63–i69.

[52] Sazawal S, Black RE, Jalla S, Mazumdar S, Sinha A, Bhan MK: Zinc supplementation reduces the incidence of acute lower respiratory infections in infants and preschool children: a double-blind, controlled trial. Pediatrics 1998;102(1 Pt 1):1–5.

[53] Das RR, Singh M: Oral zinc for the common cold. JAMA 2014;311(14):1440–1441.

[54] Turvey SE, Broide DH: Innate immunity. J Allergy Clin Immunol 2010;125(2 Suppl 2): 24–32.

[55] Bozalioğlu S, Ozkan Y, Turan M, Simşek B: Prevalence of zinc deficiency and immune response in short-term hemodialysis. J Trace Elem Med Biol 2005;18(3):243–249.

[56] Oztürk G, Erbas D, Imir T, Bor NM: Decreased natural killer (NK) cell activity in zinc-deficient rats. Gen Pharmacol 1994;25(7):1499–1503.

[57] Hujanen ES, Seppä ST, Virtanen K: Polymorphonuclear leukocyte chemotaxis induced by zinc, copper and nickel in vitro. Biochim Biophys Acta 1995;1245(2):145–152.

[58] Weston WL, Huff JC, Humbert JR, Hambidge KM, Neldner KH, Walravens PA: Zinc correction of defective chemotaxis in acrodermatitis enteropathica. Arch Dermatol 1977;113(4):422–425.

[59] Ibs KH, Rink L: Zinc-altered immune function. J Nutr 2003;133(5 Suppl 1):1452–1456.

[60] Dubben S, Hönscheid A, Winkler K, Rink L, Haase H: Cellular zinc homeostasis is a regulator in monocyte differentiation of HL-60 cells by 1 alpha,25-dihydroxyvitamin D3. J Leukoc Biol 2010;87(5):833–844.

[61] Fernandes G, Nair M, Onoe K, Tanaka T, Floyd R, Good RA: Impairment of cell-mediated immunity functions by dietary zinc deficiency in mice. Proc Natl Acad Sci U S A 1979;76(1):457–461.

[62] Allen JI, Kay NE, McClain CJ: Severe zinc deficiency in humans: association with a reversible T-lymphocyte dysfunction. Ann Intern Med 1981;95(2):154–157.

[63] Hönscheid A, Rink L, Haase H: T-lymphocytes: a target for stimulatory and inhibitory effects of zinc ions. Endocr Metab Immune Disord Drug Targets 2009;9(2):132–144.

[64] DePasquale-Jardieu P, Fraker PJ: Interference in the development of a secondary immune response in mice by zinc deprivation: persistence of effects. J Nutr 1984;114(10):1762–1769.

[65] Bao S, Liu MJ, Lee B, Besecker B, Lai JP, Guttridge DC, Knoell DL: Zinc modulates the innate immune response in vivo to polymicrobial sepsis through regulation of NF-kappaB. Am J Physiol Lung Cell Mol Physiol 2010;298(6):744–754.

[66] Knoell DL, Julian MW, Bao S, Besecker B, Macre JE, Leikauf GD, DiSilvestro RA, Crouser ED: Zinc deficiency increases organ damage and mortality in a murine model of polymicrobial sepsis. Crit Care Med 2009;37(4):1380–1388.

[67] Besecker BY, Exline MC, Hollyfield J, Phillips G, Disilvestro RA, Wewers MD, Knoell DL: A comparison of zinc metabolism, inflammation, and disease severity in critically ill infected and noninfected adults early after intensive care unit admission. Am J Clin Nutr 2011;93(6):1356–1364.

[68] Prasad AS, Bao B, Beck FW, Sarkar FH: Zinc-suppressed inflammatory cytokines by induction of A20-mediated inhibition of nuclear factor-kB. Nutrition 2011;27(7–8):816–823.

[69] Liu MJ, Bao S, Gálvez-Peralta M, Pyle CJ, Rudawsky AC, Pavlovicz RE, Killilea DW, Li C, Nebert DW, Wewers MD, Knoell DL: ZIP8 regulates host defense through zinc-mediated inhibition of NF-kB. Cell Rep 2013;3(2):386–400.

[70] Hayden MS, Ghosh S: Shared principles in NF-kappaB signaling. Cell 2008;132(3):344–362.

[71] Libermann TA, Baltimore D: Activation of interleukin-6 gene expression through the NF-kappa B transcription factor. Mol Cell Biol 1990;10(5):2327–2334.

[72] Maret W: Zinc coordination environments in proteins as redox sensors and signal transducers. Antioxid Redox Signal 2006;8(9–10):1419–1441.

[73] Bray TM, Bettger WJ: The physiological role of zinc as an antioxidant. Free Radic Biol Med 1990;8(3):281–291.

[74] Kröncke KD, Fehsel K, Schmidt T, Zenke FT, Dasting I, Wesener JR, Bettermann H, Breunig KD, Kolb-Bachofen V: Nitric oxide destroys zinc–sulfur clusters inducing zinc release from metallothionein and inhibition of the zinc finger-type yeast transcription activator LAC9. Biochem Biophys Res Commun 1994;200(2):1105–1110.

[75] Haase H, Rink L: The immune system and the impact of zinc during aging. Immun Ageing 2009;6:9.

[76] Wong CP, Ho E: Zinc and its role in age-related inflammation and immune dysfunction. Mol Nutr Food Res 2012;56(1):77–87.

[77] Markesbery WR, Ehmann WD, Alauddin M, Hossain TIM: Brain trace element concentrations in aging. Neurobiol Aging 1984;5(1):19–28.

[78] Colvin RA, Lai B, Holmes WR, Lee D: Understanding metal homeostasis in primary cultured neurons. Studies using single neuron subcellular and quantitative metallomics. Metallomics 2015;7(7):1111–1123.

[79] Pérez-Clausell J, Danscher G: Intravesicular localization of zinc in rat telencephalic boutons. A histochemical study. Brain Res 1985;337(1):91–98.

[80] Frederickson CJ, Howell GA, Haigh MD, Danscher G: Zinc-containing fiber systems in the cochlear nuclei of the rat and mouse. Hear Res 1988;36(2–3):203–211.

[81] Buxani-Rice S, Ueda F, Bradbury MW: Transport of zinc-65 at the blood–brain barrier during short cerebrovascular perfusion in the rat: its enhancement by histidine. J Neurochem 1994;62(2):665–672.

[82] Grabrucker AM: Zinc in the developing brain, in: Moran VH, Lowe N (Eds): Nutrition and the developing brain. CRC Press 2016;143–168 [in press].

[83] Inoue T, Hatayama M, Tohmonda T, Itohara S, Aruga J, Mikoshiba K: Mouse Zic5 deficiency results in neural tube defects and hypoplasia of cephalic neural crest derivatives. Dev Biol 2004;270(1):146–162.

[84] Chowanadisai W, Graham DM, Keen CL, Rucker RB, Messerli MA: Neurulation and neurite extension require the zinc transporter ZIP12 (slc39a12). Proc Natl Acad Sci U S A 2013;110(24):9903–9908.

[85] Chowanadisai W, Kelleher SL, Lönnerdal B: Maternal zinc deficiency reduces NMDA receptor expression in neonatal rat brain, which persists into early adulthood. J Neurochem 2005;94(2):510–519.

[86] Aimo L, Mackenzie GG, Keenan AH, Oteiza PI: Gestational zinc deficiency affects the regulation of transcription factors AP-1, NF-kB and NFAT in fetal brain. J Nutr Biochem 2010;21(11):1069–1075.

[87] Nuttall JR, Supasai S, Kha J, Vaeth BM, Mackenzie GG, Adamo AM, Oteiza PI: Gestational marginal zinc deficiency impaired fetal neural progenitor cell proliferation by disrupting the ERK1/2 signaling pathway. J Nutr Biochem 2015;26(11):1116–1123

[88] Fosmire GJ, al-Ubaidi YY, Sandstead HH: Some effects of postnatal zinc deficiency on developing rat brain. Pediatr Res 1975;9(2):89–93.

[89] Dvergsten CL, Fosmire GJ, Ollerich DA, Sandstead HH: Alterations in the postnatal development of the cerebellar cortex due to zinc deficiency. I. Impaired acquisition of granule cells. Brain Res 1983;271(2):217–226.

[90] Dvergsten CL, Fosmire GJ, Ollerich DA, Sandstead HH: Alterations in the postnatal development of the cerebellar cortex due to zinc deficiency. II. Impaired maturation of Purkinje cells. Brain Res 1984;318(1):11–20.

[91] Dvergsten CL, Johnson LA, Sandstead HH: Alterations in the postnatal development of the cerebellar cortex due to zinc deficiency. III. Impaired dendritic differentiation of basket and stellate cells. Brain Res 1984;318(1):21–26.

[92] Wang FD, Bian W, Kong LW, Zhao FJ, Guo JS, Jing NH: Maternal zinc deficiency impairs brain nestin expression in prenatal and postnatal mice. Cell Res 2001;11(2):135–141.

[93] Adamo AM, Zago MP, Mackenzie GG, Aimo L, Keen CL, Keenan A, Oteiza PI: The role of zinc in the modulation of neuronal proliferation and apoptosis. Neurotox Res 2010;17(1):1–14.

[94] Pang W, Leng X, Lu H, Yang H, Song N, Tan L, Jiang Y, Guo C: Depletion of intracellular zinc induces apoptosis of cultured hippocampal neurons through suppression of ERK signaling pathway and activation of caspase-3. Neurosci Lett 2013;552:140–145.

[95] Oteiza PI, Hurley LS, Lönnerdal B, Keen CL: Effects of marginal zinc deficiency on microtubule polymerization in the developing rat brain. Biol Trace Elem Res 1990;24(1): 13–23.

[96] Mackenzie GG, Salvador GA, Romero C, Keen CL, Oteiza PI: A deficit in zinc availability can cause alterations in tubulin thiol redox status in cultured neurons and in the developing fetal rat brain. Free Radic Biol Med 2011;51(2):480–489.

[97] Mackenzie GG, Oteiza PI: Zinc and the cytoskeleton in the neuronal modulation of transcription factor NFAT. J Cell Physiol 2007;210(1):246–256.

[98] Grabrucker S, Jannetti L, Eckert M, Gaub S, Chhabra R, Pfaender S, Mangus K, Reddy PP, Rankovic V, Schmeisser MJ, Kreutz MR, Ehret G, Boeckers TM, Grabrucker AM: Zinc deficiency dysregulates the synaptic ProSAP/Shank scaffold and might contribute to autism spectrum disorders. Brain 2014;137(Pt1):137–152.

[99] Peters DP: Effects of prenatal nutritional deficiency on affiliation and aggression in rats. Physiol Behav 1978;20(4):359–362.

[100] Halas ES, Hanlon MJ, Sandstead HH: Intrauterine nutrition and aggression. Nature 1975;257(5523):221–222.

[101] Sandstead HH, Fosmire GJ, Halas ES, Jacob RA, Strobel DA, Marks EO: Zinc deficiency: effects on brain and behavior of rats and rhesus monkeys. Teratology 1977;16(2): 229–234.

[102] Golub MS, Gershwin ME, Vijayan VK: Passive avoidance performance of mice fed marginally or severely zinc deficient diets during post-embryonic brain development. Physiol Behav 1983;30(3):409–413.

[103] Halas ES, Eberhardt MJ, Diers MA, Sandstead HH: Learning and memory impairment in adult rats due to severe zinc deficiency during lactation. Physiol Behav 1983;30(3):371–381.

[104] Halas ES, Hunt CD, Eberhardt MJ: Learning and memory disabilities in young adult rats from mildly zinc deficient dams. Physiol Behav 1986;37(3):451–458.

[105] Tahmasebi Boroujeni S, Naghdi N, Shahbazi M, Farrokhi A, Bagherzadeh F, Kazemnejad A, Javadian M: The effect of severe zinc deficiency and zinc supplement on spatial learning and memory. Biol Trace Elem Res 2009;130(1):48–61.

[106] Lokken PM, Halas ES, Sandstead HH: Influence of zinc deficiency on behavior. Proc Soc Exp Biol Med 1973;144(2):680–682.

[107] Halas ES, Sandstead HH: Some effects of prenatal zinc deficiency on behavior of the adult rat. Pediatr Res 1975;9(2):94–97.

[108] Holst B, Egerod KL, Schild E, Vickers SP, Cheetham S, Gerlach LO, Storjohann L, Stidsen CE, Jones R, Beck-Sickinger AG, Schwartz TW: GPR39 signaling is stimulated by zinc ions but not by obestatin. Endocrinology 2007;148(1):13–20.

[109] Młyniec K, Gaweł M, Librowski T, Reczyński W, Bystrowska B, Holst B: Investigation of the GPR39 zinc receptor following inhibition of monoaminergic neurotransmission and potentialization of glutamatergic neurotransmission. Brain Res Bull 2015;115:23–29.

[110] Pochwat B, Nowak G, Szewczyk B: Relationship between zinc (Zn (2+)) and glutamate receptors in the processes underlying neurodegeneration. Neural Plast 2015;2015:591563.

[111] Bresink I, Ebert B, Parsons CG, Mutschler E: Zinc changes AMPA receptor properties: results of binding studies and patch clamp recordings. Neuropharmacology 1996;35(4): 503–509.

[112] Kalappa BI, Anderson CT, Goldberg JM, Lippard SJ, Tzounopoulos T: AMPA receptor inhibition by synaptically released zinc. Proc Natl Acad Sci U S A 2015;112(51): 15749–15754.

[113] Grabrucker AM, Knight MJ, Proepper C, Bockmann J, Joubert M, Rowan M, Nienhaus GU, Garner CC, Bowie JU, Kreutz MR, Gundelfinger ED, Boeckers TM: Concerted action of zinc and ProSAP/Shank in synaptogenesis and synapse maturation. EMBO J 2011;30(3):569–581.

[114] Yu X, Ren T, Yu X: Disruption of calmodulin-dependent protein kinase II α/brain-derived neurotrophic factor (α-CaMKII/BDNF) signalling is associated with zinc deficiency-induced impairments in cognitive and synaptic plasticity. Br J Nutr 2013;110(12):2194–2200.

[115] Manosso LM, Moretti M, Ribeiro CM, Gonçalves FM, Leal RB, Rodrigues AL: Antidepressant-like effect of zinc is dependent on signaling pathways implicated in BDNF modulation. Prog Neuropsychopharmacol Biol Psychiatry 2015;59:59–67.

[116] Takeda A, Tamano H: Significance of the degree of synaptic Zn2+ signaling in cognition. Biometals 2015. [Epub ahead of print].

[117] Minami A, Sakurada N, Fuke S, Kikuchi K, Nagano T, Oku N, Takeda A: Inhibition of presynaptic activity by zinc released from mossy fiber terminals during tetanic stimulation. J Neurosci Res 2006;83(1):167–176.

[118] Chappell RL, Anastassov I, Lugo P, Ripps H: Zinc-mediated feedback at the synaptic terminals of vertebrate photoreceptors. Exp Eye Res 2008;87(4):394–397.

[119] Keller KA, Grider A, Coffield JA: Age-dependent influence of dietary zinc restriction on short-term memory in male rats. Physiol Behav 2001;72(3):339–348.

[120] Whittle N, Hauschild M, Lubec G, Holmes A, Singewald N: Rescue of impaired fear extinction and normalization of cortico-amygdala circuit dysfunction in a genetic mouse model by dietary zinc restriction. J Neurosci 2010;30(41):13586–13596.

[121] Gao HL, Xu H, Xin N, Zheng W, Chi ZH, Wang ZY: Disruption of the CaMKII/CREB signaling is associated with zinc deficiency-induced learning and memory impairments. Neurotox Res 2011;19(4):584–591.

[122] Daumas S, Halley H, Lassalle JM: Disruption of hippocampal CA3 network: effects on episodic-like memory processing in C57BL/6J mice. Eur J Neurosci 2004;20(2):597–600.

[123] Takeda A, Tamano H, Kan F, Hanajima T, Yamada K, Oku N: Enhancement of social isolation-induced aggressive behavior of young mice by zinc deficiency. Life Sci 2008;82(17–18):909–914.

[124] Tassabehji NM, Corniola RS, Alshingiti A, Levenson CW: Zinc deficiency induces depression-like symptoms in adult rats. Physiol Behav 2008;95(3):365–369.

[125] O'Dell BL: Roles of zinc and copper in the nervous system. Prog Clin Biol Res 1993;380:147–162.

[126] Prasad AS, Fitzgerald JT, Hess JW, Kaplan J, Pelen F, Dardenne M: Zinc deficiency in elderly patients. Nutrition 1993;9(3):218–224.

[127] Pepersack T, Rotsaert P, Benoit F, Willems D, Fuss M, Bourdoux P, Duchateau J: Prevalence of zinc deficiency and its clinical relevance among hospitalised elderly. Arch Gerontol Geriatr 2001;33(3):243–253.

[128] Mocchegiani E, Romeo J, Malavolta M, Costarelli L, Giacconi R, Diaz LE, Marcos A: Zinc: dietary intake and impact of supplementation on immune function in elderly. Age (Dordr) 2013;35(3):839–860.

[129] Nuttall JR, Oteiza PI: Zinc and the aging brain. Genes Nutr 2014;9(1):379.

[130] Takeda A: Insight into glutamate excitotoxicity from synaptic zinc homeostasis. Int J Alzheimers Dis 2010;2011:491597.

[131] Saito T, Takahashi K, Nakagawa N, Hosokawa T, Kurasaki M, Yamanoshita O, Yamamoto Y, Sasaki H, Nagashima K, Fujita H: Deficiencies of hippocampal Zn and ZnT3 accelerate brain aging of rat. Biochem Biophys Res Commun 2000;279(2):505–511.

[132] Walsh CT, Sandstead HH, Prasad AS, Newberne PM, Fraker PJ: Zinc: health effects and research priorities for the 1990s. Environ Health Perspect 1994;102(Suppl 2):5–46.

[133] Prasad AS, Halsted JA, Nadimi M: Syndrome of iron deficiency anemia, hepatosplenomegaly, hypogonadism, dwarfism and geophagia . J Trace Elem Exp Med 2001;14:121–144.

[134] Sandstead HH, Prasad AS, Schulert AR, Farid Z, Miale A Jr, Bassilly S, Darby W J: Human zinc deficiency, endocrine manifestations and response to treatment. Am J Clin Nutr 1967;20:422–442.

[135] Pfaender S, Grabrucker AM: Characterization of biometal – profiles in neurological disorders. Metallomics 2014;6(5):960–977.

[136] Prakash A, Bharti K, Majeed AB: Zinc: indications in brain disorders. Fundam Clin Pharmacol 2015;29:131–149.

[137] American Psychiatric Association. Diagnostic and Statistical Manual of Mental Disorders, 5th edition. Arlington, VA: American Psychiatric Publishing, 2013.

[138] Grabrucker AM: Environmental factors in autism. Front Psychiatry 2012;3:118.

[139] Grabrucker AM: A role for synaptic zinc in ProSAP/Shank PSD scaffold malformation in autism spectrum disorders. Dev Neurobiol 2014;74(2):136–146.

[140] Yasuda H, Yoshida K, Yasuda Y, Tsutsui T: Infantile zinc deficiency: association with autism spectrum disorders. Sci Rep 2011;1:129.

[141] Bjørklund G: The role of zinc and copper in autism spectrum disorders. Acta Neurobiol Exp 2013;73:225–236.

[142] Hagmeyer S, Haderspeck J, Grabrucker AM: Behavioral impairments in animal models for zinc deficiency. Front Behav Neurosci 2015;8:443.

[143] Barnham KJ, Bush AI: Metals in Alzheimer's and Parkinson's diseases. Curr Opin Chem Biol 2008;12:222–228.

[144] Szewczyk B: Zinc homeostasis and neurodegenerative disorders. Front Aging Neurosci 2013;5:33.

[145] Ling Y, Morgan K, Kalsheker N: Amyloid precursor protein (APP) and the biology of proteolytic processing: relevance to Alzheimer's disease. Int J Biochem Cell Biol 2003;35:1505–1535.

[146] Wilquet V, De SB: Amyloid-beta precursor protein processing in neurodegeneration. Curr Opin Neurobiol 2004;14:582–588.

[147] Bush AI, Pettingell WH Jr, Paradis MD, Tanzi RE: Modulation of A beta adhesiveness and secretase site cleavage by zinc. J Biol Chem 1994;269:12152–12158.

[148] Grabrucker AM, Schmeisser MJ, Udvardi PT, Arons M, Schoen M, Woodling NS, Andreasson KI, Hof PR, Buxbaum JD, Garner CC, Boeckers TM: Amyloid beta protein-induced zinc sequestration leads to synaptic loss via dysregulation of the ProSAP2/Shank3 scaffold. Mol Neurodegener 2011;6:65.

[149] Adlard PA, Parncutt JM, Finkelstein DI, Bush AI: Cognitive loss in zinc transporter-3 knock-out mice: a phenocopy for the synaptic and memory deficits of Alzheimer's disease? J Neurosci 2010;30(5):1631–1636.

[150] Vallee BL, Falchuk KH: The biochemical basis of zinc physiology. Physiol Rev 1993;73(1):79–118.

[151] Nygaard SB, Larsen A, Knuhtsen A, Rungby J, Smidt K: Effects of zinc supplementation and zinc chelation on in vitro β-cell function in INS-1E cells. BMC Res Notes 2014;7:84.

[152] Taylor CG: Zinc, the pancreas, and diabetes: insights from rodent studies and future directions. Biometals 2005;18(4):305–312.

[153] Kinlaw WB, Levine AS, Morley JE, Silvis SE, McClain CJ: Abnormal zinc metabolism in type II diabetes mellitus. Am J Med 1983;75(2):273–277.

[154] Al-Maroof RA, Al-Sharbatti SS: Serum zinc levels in diabetic patients and effect of zinc supplementation on glycemic control of type 2 diabetics. Saudi Med J 2006;27(3):344–350.

[155] Simon SF, Taylor CG: Dietary zinc supplementation attenuates hyperglycemia in db/db mice. Exp Biol Med 2001;226(1):43–51.

[156] Jayawardena R, Ranasinghe P, Galappatthy P, Malkanthi R, Constantine G, Katulanda P: Effects of zinc supplementation on diabetes mellitus: a systematic review and meta-analysis. Diabetol Metab Syndr 2012;4(1):13

[157] Ranasinghe P, Pigera S, Galappatthy P, Katulanda P, Constantine GR: Zinc and diabetes mellitus: understanding molecular mechanisms and clinical implications. DARU J Pharm Sci 2015;23:44.

[158] Kaur K, Gupta R, Saraf SA, Saraf SK: Zinc: The Metal of Life. Comprehensive Reviews in Food Science and Food Safety,2014;13:358–376.

[159] Ayton S, Lei P, Bush AI. Biometals and their therapeutic implications in Alzheimer's disease. Neurotherapeutics 2015; 12(1):109–120.

[160] Lee S, Zheng X, Krishnamoorthy J, Savelieff MG, Park HM, Brender JR, Kim JH, Derrick JS, Kochi A, Lee HJ, Kim C, Ramamoorthy A, Bowers MT, Lim MH. Rational design of a structural framework with potential use to develop chemical reagents that target and modulate multiple facets of Alzheimer's disease. J Am Chem Soc 2014; 136(1): 299–310.

Chapter 3

Zinc: What Is Its Role in Lung Cancer?

Nidia N. Gomez, Verónica S. Biaggio,
María E. Ciminari, María V. Pérez Chaca and
Silvina M. Álvarez

Additional information is available at the end of the chapter

http://dx.doi.org/10.5772/63209

Abstract

Recently, zinc emerged as an important signaling molecule, activating intracellular pathways and regulating cell fate, although our knowledge remains incomplete. Zinc is required in many enzymatic and metabolic pathways, playing roles as enzyme cofactors. In normal cell physiology, optimal zinc availability is essential for regular growth and proliferation. Zinc accumulation has varied effects: from stimulation to inhibition of cell growth, depending on type. There is evidence that zinc is capable of inducing apoptosis in some cancers, while others proved that zinc may act as apoptosis activator depending on the dose and cell type. Upregulation of telomerase in most cancer tissues is considered to be responsible for unlimited proliferation of cancer cells, and in some cell lines, it was induced by Zn. These suggest that Zn is highly involved in cell cycle and metabolism; whether it goes to the survival or the cancer pathway depends on the concentration and the cell type involved. Nevertheless, the conclusion is that Zn is not just another trace element; but a vital one and further studies are needed to elucidate the mechanisms involved in cancer and metastatic spread in order to identify potential therapies.

Keywords: functions of zinc, zinc deficiency, high concentration of zinc, cancer, zinc, lung cancer, deficiency, homeostasis

1. Introduction

Zinc is one of the most important trace elements in the body. It has a catalytic/regulatory role in many enzymes, maintains the structural integrity of different proteins, and modulates protein-protein interactions. At the cellular level, zinc is essential for cell proliferation and survival, contributes to genomic stability and antioxidant defense, which highlights its crucial role in aging and age-dependent degenerative diseases. Zinc is indispensable for proper immune function, its insufficiency may exacerbate immune-senescence, and zinc supplementation is beneficial to immune responses in the elderly [1]. In the intracellular environment, zinc interacts with signal transducers implicated in immune response and influences both the structural stability and function of immunologically relevant transcription factors [2]. Zinc is needed for DNA synthesis, RNA transcription, cell division, and cell activation. Programmed cell death (apoptosis) is potentiated in the absence of adequate levels of zinc under physiological conditions [3].

During the past four decades, a spectrum of zinc clinical deficiencies in human subjects has emerged. On one hand, the manifestations of zinc deficiency (ZD) may be severe, and on the other end of the spectrum, ZD may be mild or marginal [3]. Micronutrient deficiencies are an important and global public health problem. In Mexico, the first comprehensive picture of the frequency and distribution of micronutrient deficiencies was presented by the Mexican National Nutrition Survey of 1999. Zinc deficiency was the second most common micronutrient deficiency; 34% of which was found in infants [4]. Several studies have now confirmed that ZD is fairly prevalent in developing countries, affecting nearly two billion subjects and that growth retardation commonly observed in these countries may indeed be due to ZD [5]. Increased prevalence of obstructive lung disorders and lung cancer is associated with low dietary Zinc (Zn) intake and thought to be due, at least in part, to protective effects of Zn against cadmium, which is toxic and accumulates in alveolar macrophages (AM). Among the actions of Zn ions on the immune system are its effects on phagocytic cells. Hamon and colleagues [6] suggested links between lung injury, impaired phagocytosis, and Zn deficiency. In the bloodstream, zinc insufficiency may also contribute to cardiovascular risk via its association with reduced antioxidant capacity [7], endothelial dysfunction [8], arterial wall stiffness, and increased systolic blood pressure [9, 10].

The use of zinc as a nutritional supplement has become very common in many countries. However, adverse effects of high doses of zinc supplement on immunologic functions have not been clearly defined. Some studies from animal models show that high dietary zinc increased the functions of T lymphocytes and macrophages [11, 12]. Other studies reported a decrease in lymphocyte stimulation response, chemotaxis and bacterial phagocytosis of polymorphonuclear leukocytes, and monocyte function and neutropenia by high oral intake of zinc [13, 14].

2. Zinc, extracellular matrix, and cancer

Zinc was demonstrated to have the ability to neutralize free radicals protecting the body from harmful effects, immune disorders, and increased risk of cancer [15]; therefore, the deficiency of zinc may increase oxidative stress [16, 17]. Importantly, an increase of oxidative and nitrosative stress and inflammation in rat lung, in a stage of marginal zinc deficiency, was found [16, 18].

Cell-cell and cell-extracellular matrix interactions are essential in the development and maintenance of normal tissue cytoarchitecture and play an important role in the development and progression of many types of cancer [19]. Simultaneously, with the changes causing the immortalization of epithelial cells, there is a gradual evolution of the tumor microenvironment [19].

The extracellular matrix (ECM) represents a very complex network of structurally, mechanically, and biochemically heterogeneous components [20] including: collagen, elastin, fibrillin, fibulin, glycoproteins, and integrin receptors of ECM components [19, 21]. The systems that regulate deposition and stability of the ECM also include chaperones and enzymes that catalyze the post-translational processing of ECM components, as well as systems that destabilize and degrade the ECM to facilitate its renewal [20, 21].

It is known that matrix metalloproteinases (MMPs) are a family of zinc dependent endopeptidases which main function is to degrade and deposit structural proteins within the ECM. The production of MMPs is stimulated by factors such as oxidative stress, growth factors, and inflammation which lead to its up or downregulation with subsequent ECM remodeling [22]. Normally, excess MMPs activation is controlled by tissue inhibitors of metalloproteinases (TIMPs). MMPs and TIMPs imbalance has been implicated in multiple diseases [22]. Recent studies have demonstrated that ECM and basement membrane degradation by MMPs play an important role in tumorigenesis by modulating cell proliferation, apoptosis, and host immune surveillance, tumor invasion and metastasis [23]. In addition, MMP-9 acts as an important oncogene, thereby improving the invasiveness of cancer cells. It has been suggested that a high level of MMP-9 confers a poor prognosis in various cancers [24]. On the other hand, MMP2 (gelatinase A) has attracted particular interest in neoplasias, since it degrades type IV collagen, a major component of basement membrane undergoing destruction at an early stage of the invasive process [25].

It is generally accepted that a fundamental process for distant metastasis formation comprises epithelial-mesenchymal transition (EMT), during which tumor cells lose their epithelial properties and acquire a fibroblast-like phenotype; as a consequence, reduced intercellular adhesion, enhanced invasiveness, and increased apoptotic resistance of cells [26, 27]. In the early stages of tumor, several signaling pathways are activated, such as growth factors and zinc-finger transcription factors including Snail [26, 28]. In fact, Snail-1 has been shown to be crucial during cancer progression and metastasis. In colon cancer patients, enhanced levels of Snail1 are usually associated with poor clinical outcome [26, 29]. Recent studies in Snail-1 deficient mouse embryos support the idea that transcriptional repression of E-cadherin (cellular adhesion molecules) is associated with Snail-1 activity [26].

Marginal ZD was also associated with oxidative stress and inflammation in both mammary gland ductal epithelium and adipocyte-rich stromal. Excess of collagen directly inhibits mammary gland expansion and has major implications for breast disease risk [30]. Zinc-dependent enzyme MMP2 activity, a critical protein for ductal elongation and infiltration into the mammary fat pad, was also modified in ZD mice mammary glands [30].

Heat shock proteins are chaperones that play a pivotal role in cells survival under stressful conditions. Under normal conditions, Heat shock protein 27(Hsp27) is weakly expressed in cells; however, once stress occurs, Hsp27 expression increases, exerting an anti-oxidative damage function [31]. We analyzed the effect of Zn deficiency on the expression of cytoprotective factors (Hsp27 and Hsp70i) where both chaperones increased and are consistently associated with cellular stress and inflammation in lung [32, 33]. Likewise, Hsp70 has a role in iNOS induction [16, 31, 33]. In addition, proliferating cell nuclear antigen (PCNA) expression was increased in the ZD group [31, 34]. Qin and colleagues [31] demonstrated a possible association between PCNA and Hsp27 expression in retinoblastoma tumor. There is a lot of evidence that PCNA expression can be used as an index for evaluating malignant tumor cells proliferation, as well as the malignant potential and prognosis of a tumor [31, 35, 36].

Molecular mechanisms that define the pathological and physiological activities of EMT in distinct cellular contexts likely intersect. Zinc could act as trigger factor of events cascade in the ECM. Therefore, in order to understand some of the pathological mechanisms involved in cancer and metastasis, it would be necessary for more studies.

3. Zinc and apoptosis in lung cancer cells: excess or default?

Zinc plays a role in several intracellular signaling pathways, and its deregulation is present in various cancers. Levels of zinc in serum and malignant tissues of patients with various types of cancer are abnormal, supporting the involvement of zinc in cancer development. Imbalance of zinc transporters cause intracellular and serum zinc levels alteration. Patients with lung, breast, liver, and prostate cancer exhibit zinc deregulation in a meta-analysis performed in [37]. Zinc level decreases in lung cancer, but it is unclear whether hypozincaemia is a consequence of tumor, chronic stress, or a combination of both effects [38]. Stress, infection, or chronic diseases lead to redistribution of zinc between body compartments, and thus reduce zincaemia [39].

Zinc is required for both normal cell survival and for cell death via its role in apoptosis, which is strongly regulated. Its deregulation is central to the pathogenesis of a number of diseases, including cancer. As such, the factors like zinc that regulates the execution phases of apoptosis are of great interest as potential therapies. Free zinc ion does not only act as an inhibitor but also may act as an activator of apoptosis depending on dose and cell type. Franklin et al. [40] reviewed the effects of zinc on the regulation of apoptosis in malignant cells, because it is reported to both induce apoptosis in some cancers and to protect other cancer cells against apoptosis induced by other factors. They studied prostate, breast, liver, pancreas, and ovarian cancer finding that zinc is an apoptogenic agent in ovarian epithelial cells, in breast epithelial cells, and in prostate cells [40].

It is well known that lung cancer is the leading cause of cancer-related deaths in both men and women. Lung cancer is subdivided into two types based on cell type and pathology: small-cell lung cancers and non-small-cell lung cancers (NSCLCs), of which approximately 85% are NSCLCs [41]. At the advanced stages, taxane chemotherapy regimens are commonly used for the treatment of NSCLCs as first-line options, but the therapeutics results are not satisfactory. Kocdor et al. [42] found that zinc exhibited growth inhibitory and apoptotic effects in a dose-dependent manner, up to the IC_{50} concentrations for cultured lung cancer cells. Importantly, these effects were significantly increased when zinc and docetaxel (derived from the paclitaxel—natural compound isolated from pacific yew tree bark) were combined to treat lung cancer. Importantly, zinc deficiency reduces paclitaxel efficacy in cultured prostate cancer cells, whereas increased intracellular zinc concentrations sensitize prostate cancer cells to cytotoxic agents, including paclitaxel, via inhibition of NF-κB activation [43, 44]. Therefore, authors proposed that zinc supplementation may have growth inhibitory effects against NSCLC cells and may increase docetaxel efficacy [42]. The semisynthetic form, docetaxel, primarily stabilizes cytoplasmic microtubules via binding to the β-tubulin site, causing cell cycle arrest at the G2/M phase and driving apoptosis. Therefore, the IC_{50} doses of docetaxel and zinc were higher for the p53-null H1299 cells than A549 cells. Functional p53 status may influence docetaxel and zinc-induced cytotoxicity. PTEN-PI3K-Akt-Bax signaling cascade is involved in the therapeutic effect of combined radiation/paclitaxel treatment in NSCLC without p53 expression [45].

John et al. [46] proposed that zinc depletion induces cell death via apoptosis (or necrosis if apoptotic pathways are blocked), while sufficient zinc levels allows maintenance of cell survival pathways such as autophagy and regulation of reactive oxygen species. Although in the results shown by meta-analysis zinc tissue levels are low in lung cancer [37, 38], two older studies demonstrated the contrary: they exhibit elevated zinc levels when compared with the corresponding normal tissues [47, 48]. Interestingly, while data of zinc levels in tumor tissue is limited, it has been widely recognized that ZIP is upregulated in most cancers, thereby indicating increased zinc concentrations in tumor majority [46]. Additionally, peripheral tissue surrounding lung metastasis has higher zinc content than the corresponding normal tissue or the tumor tissue itself [47]. Consequently, zinc levels regulation to promote immune cells survival and tumor apoptosis are in order. Likewise adjustments in zinc homeostasis may be a contributing factor in genetic alterations (ZNT, ZIP, metallothionein, etc.) or environmental causes (nutritional status, exposure to zinc, microbial control) playing a role in the genesis and/or maintenance of cancer [46].

Lung cancer chemotherapy treatments itself do not produce satisfactory results; however, apoptogenic effect of zinc increased docetaxel therapy efficacy against NSCLC, achieving good results. Zinc transporters are upregulated in lung cancer, but more data are needed to clarify zinc tissue tumor values and its role in the triggering and progress of apoptosis to finally found a successful therapy facing this scourge.

4. Zinc and telomerase

A telomere is a repetitive sequence of DNA that protects the ends of linear chromosomes from deterioration and repair activity [49]. Mammalian telomeres consist of repetitive TTAGGG sequences that are crucial to formation of the capping structures, which are bound by telomere-binding factors called shelterin [50]. The shelterin complex is a six subunit complex composed of directly binding proteins TRF1, TRF2, and POT16 and their associated proteins Rap1, TPP1, and TIN2 [51].

Due to the inability of DNA polymerase to replicate the 5′ end of the lagging strand, the length of telomeres is shortened progressively with each cell division, which eventually leads to cellular senescence when the telomere length is reduced beyond the critical level [52]. It is believed that the maintenance of telomeres is essential for the immortality of cancer cells. Telomeres are maintained by a specialized reverse transcriptase, the ribonucleoprotein telomerase, which is composed of a ubiquitously expressed RNA subunit, human telomerase RNA component [hTERC; 53], and a protein catalytic subunit, human telomerase reverse transcriptase (hTERT), the expression of which is highly regulated [54].

Zn deficiency is known to suppress the proliferation of tumor cells [55], suggesting that Zn has an important role in cell proliferation. The involvement of Zn in the proliferation of lymphocytes and other non-cancer cells also has been documented [7]. But other evidence shows that tumor size has a reverse relationship with zinc amount [56, 57].

In 2000, Nemoto et al. [58] studied how zinc modulates telomerase activity, showing that treatment with 100 uM Zn enhanced telomerase activity in renal cell carcinoma (NRC-12) and human prostatic cancer (DU145) cells. This enhancing effect of Zn suggests that it may not be caused by cytotoxicity but rather by some biological events such as induction of Zn/binding proteins or activation of transcription factors containing Zn-finger motifs. Based on that, Zarghami et al. [59] studied the relation between Zn plasma and telomerase activity in bladder cancer patients. Nevertheless, they only found a significant relationship between Zn and Telomerase activity in the female patients, where the cancerous patients presented less Zn concentrations than the control patients, with elevated enzyme activity, consistent with the findings of Whelan et al. [60]. The study of Prasad et al. [61] also showed that patients with head and neck cancer presenting bigger tumors had zinc deficiency as well [61]. Similar results were found in lung cancer patients [62]. Therefore, there are paradoxical results regarding zinc levels and its effect on cancer.

More recently, a new embryonic stem cell marker was discovered, zinc finger and SCAN domain containing four genes (Zscan4), which has a key function in genomic stability by regulating telomere elongation, and might also have a fundamental role in the mechanism controlling telomere length regulation [63]. Zscan was also found to promote telomere elongation during reprogramming, but it is not associated with increased telomerase activity [64]. It has been shown that overexpression of Zscan4 rescues cell proliferation and causes rapid telomere extension [64]. In 2014, Lee and Gollahon [65] showed that Zscan4 binds directly to the shelterin complex member Rap1. Apparently, the binding between Zscan4 and Rap1 may be required for disrupting telomere protection dissociation of the t-loop to control telomere length in telomere biology of cancer cells.

Another zinc-finger protein involved in cell differentiation, senescence, and apoptosis is Zfp637. It belongs to the Kruppel-like protein family and comprises six consecutively typical and one atypical C2H2 zinc-finger motifs. It has been reported that Zfp637 is located in nucleus and behaves as a repression regulator in myogenic cellular differentiation by promoting mTERT expression [66]. Recently, in [67], it was provided the first mechanism through which Zfp637 protects cells against oxidative stress-induced premature senescence. Zfp637 binds to the mTERT promoter and transcriptionally activates mTERT. mTERT expression maintains telomerase activity and telomere length and promotes cell proliferation. On the other hand, the oxidative stress-triggered downregulation of Zfp637 results in depressed binding of Zfp637 to the mTERT promoter, leading to reduced levels of mTERT-dependent telomerase activity and accelerated telomere shortening and cellular senescence, what can be reverted by overexpression of Zfp637 [67].

All these studies show that several proteins involved in the activity of telomerase have Zn-finger motifs, indirectly suggesting the involvement of this ion in cancer outcome. Unfortunately, more studies need to be done in order to assure if Zn presence or deficiency the responsible for cancer onset.

5. Conclusion

The aim of this review was to look through the state-of-the-art concerning the zinc homeostasis and cancer. Zn microenvironment may play a key role in oxidative stress, apoptosis, and/or cell signaling alterations which influences the behavior of malignant cancer cells. In fact, the study of cancer biology has mainly focused on malignant epithelial cancer cells, although tumors also contain a stromal compartment, composed by different type of cells and also includes various types of macromolecules comprising the extracellular matrix. Following this rationale, several hundred zinc supplementation studies have been conducted, investigating the effects of zinc on cancer, often with contradictory results. The mechanisms responsible for Zn accumulation and the consequence of Zn dysregulation are poorly understood, and mostly dependent of the type of cell or tissue compromised. For this reason, further studies are needed to elucidate the mechanism of this protection.

Acknowledgements

The authors would like to acknowledge the financial support of PROICO 2-1814, C y T UNSL. The authors sincerely apologize many colleagues whose work they were unable to cite owing to space limitations.

Author details

Nidia N. Gomez[1,2*], Verónica S. Biaggio[1,2], María E. Ciminari[1], María V. Pérez Chaca[1] and Silvina M. Álvarez[1,2]

*Address all correspondence to: ngomez@unsl.edu.ar

1 Department of Biochemistry and Biological Sciences, Faculty of Chemistry, Biochemistry and Pharmacy, National University of San Luis, San Luis, Argentina

2 IMIBIO-CONICET, San Luis, Argentina

References

[1] Putics A, Vödrös D, Malavolta M, Mocchegiani E, Csermely P, Soti C. Zinc supplementation boosts the stress response in the elderly: Hsp70 status is linked to zinc availability in peripheral lymphocytes. Exp Gerontol. 2008;43:452–461. doi:10.1016/j.exger.2008.01.002

[2] Wellinghausen N, Rink L. The significance of zinc for leukocyte biology. J Leukoc Biol. 2008;64:571–577.

[3] Prasad AS. Clinical, immunological, anti-inflammatory and antioxidant roles of zinc. Exp Gerontol. 2008;43:370–377.

[4] Shamah-Levy T, Villalpando S, Jáuregui A, Rivera JA. Overview of the nutritional status of selected micronutrients in Mexican children in 2006. Salud Pública de México. 2012;54(2):146–151.

[5] Brown KH, Peerson JM, Allen LH, Rivera J. Effect of supplemental zinc on the growth and serum zinc concentrations of prepubertal children: a meta-analysis of randomized, controlled trials. Am J Clin Nutr. 2002;75:1062–1071.

[6] Hamon R, Homan CC, Tran HB, Mukaro VR, Lester SE, Roscioli E, Bosco MD, Murgia ChM, Ackland ML, Jersmann HP, Lang C, Zalewski PD, Hodge SJ. Zinc and zinc transporters in macrophages and their roles in efferocytosis in COPD. Plos One. 2014;9(10):e110056. doi:10.1371/journal.pone.0110056

[7] Shankar AH, Prasad AS. Zinc and immune function: the biological basis of altered resistance to infection. Am J Clin Nutr. 1998;68(2 Suppl):447S–463S.

[8] Hennig B, Meerarani P, Ramadass P, Toborek M, Malecki A, Slim R, McClain CJ. Zinc nutrition and apoptosis of vascular endothelial cells: implications in atherosclerosis. Nutrition. 1999;15:744–748. doi:10.1016/S0899-9007(99)00148-3

[9] Khadilkar AV, Chiplonkar SA, Pandit DS, Kinare AS, Khadilkar VV. Metabolic risk factors and arterial stiffness in Indian children of parents with metabolic syndrome. J Am Coll Nutr. 2012;31:54–62.

[10] De Paula RDS, Aneni EC, Ana Paula R. Costa APR, Figueiredo VN, Moura FA, Freitas WM, Quaglia LA, Santos SN, Soares AA, Nadruz W, Jr., Blaha M, Blumenthal R, Agatston A, Nasir K, Sposito AC. Low zinc levels is associated with increased inflam-

matory activity but not with atherosclerosis, arteriosclerosis or endothelial dysfunction among the very elderly. BBA Clin. 2014;2:1–6. doi:org/10.1016/j.bbacli.2014.07.002

[11] Salvin SB, Horecker BL, Pan LX, Rabin BS, The effect of dietary zinc and prothymosina on cellular immune responses of RF/J mice. Clin Immunopathol. 1987;43:281–288. doi:10.1016/0090-1229(87)90137-1

[12] Singh KP, Zaidi SI, Raisuddin S, Saxena AK, Murthy RC, Ray PK. Effect of zinc on immune functions and host resistance against infection and tumor challenge. Immunopharmacol Immunotoxicol. 1992;14(4):813–840.

[13] Chandra RK. Excessive intake of zinc impairs immune response. JAMA. 1984;525:1443–1446. doi:10.1001/jama.1984.03350110043027

[14] Fosmire GJ. Zinc toxicity. Am J Clin Nutr 1978;51:225–227.

[15] Grigorescu R, Gruia MI, Nacea V, Nitu C. Parameters of oxidative stress variation depending on the concentration of inorganic zinc compounds. J Med Life. 2015;8(4): 449–451.

[16] Gomez NN, Davicino RC, Biaggio VS, Bianco GA, Alvarez SM, Fischer P, Masnatta L, Rabinovich GA, Gimenez MS. Overexpression of inducible nitric oxide synthase and cyclooxygenase-2 in rat zinc-deficient lung: involvement of a NF-kappaB dependent pathway. Nitric Oxide. 2006;14(1):30–8. doi:10.1016/j.niox.2005.09.001

[17] Song Y, Leonard SW, Traber MG, Ho E. Zn deficiency affects DNA damage, oxidative stress, antioxidant defenses, and DNA repair in rats. J Nutr. 2009;139:1626–1631. doi:10.3945/jn.109.106369

[18] Biaggio VS, Pérez Chaca MV, Valdéz SR, Gómez NN, Gimenez MS. Alteration in the expression of inflammatory parameters as a result of oxidative stress produced by moderate zinc deficiency in rat lung. Exp Lung Res. 2010;36(1):31–44. doi: 10.3109/01902140903061787

[19] Davies KJ. The complex interaction of matrix metalloproteinases in the migration of cancer cells through breast tissue stroma. Int J Breast Cancer. 2014;2014:839094. doi: 10.1155/2014/839094

[20] van der Horst G, Bos L, van der Pluijm G. Epithelial plasticity, cancer stem cells, and the tumor-supportive stroma in bladder carcinoma. Mol Cancer Res. 2012;10(8):995–1009. doi:10.1158/1541-7786

[21] Mižíková I, Morty RE. The extracellular matrix in bronchopulmonary dysplasia: target and source. Front Med (Lausanne). 2015;2:91. doi:10.3389/fmed.2015.00091

[22] Amin M, Pushpakumar S, Muradashvili N, Kundu S, Tyagi SC, Sen U. Regulation and involvement of matrix metalloproteinases in vascular diseases. Front Biosci (Landmark Ed) 2016;21:89–118. doi:10.2741/4378

[23] Verma S, Kesh K, Gupta A, Swarnakar S. An overview of matrix metalloproteinase 9 polymorphism and gastric cancer risk. Asian Pac J Cancer Prev. 2015;16(17):7393–7400. doi:10.7314/APJCP.2015.16.17.7393

[24] Ruhul Amin ARM, Senga T, Oo ML, Thant AA, Hamaguchi M. Secretion of matrix metalloproteinase-9 by the proinflammatory cytokine, IL-1beta: a role for the dual signalling pathways, Akt and Erk. Genes Cells. 2003;6:515–523. doi:10.1046/j.1365-2443.2003.00652.x

[25] Han YH, Gao B, Huang JH, Z, Guo Z, Jie Q, Yang L, Luo ZL. Expression of CD147, PCNA, VEGF, MMPs and their clinical significance in the giant cell tumor of bones. Int J Clin Exp Pathol. 2015;8(7):8446–8452. eCollection 2015.

[26] Brzozowa M, Michalski M, Wyrobiec G, Piecuch A, Dittfeld A, Harabin-Słowińska M, Boroń D, Wojnicz R. The role of Snail1 transcription factor in colorectal cancer progression and metastasis. Contemp Oncol (Pozn) 2015;19(4):265–270. doi:10.5114/wo.2014.42173

[27] Trimboli AJ, Fukino K, de Bruin A, Wei G, Shen L, Tanner SM, Creasap N, Rosol TJ, Robinson ML, Eng C, Ostrowski MC, Leone G. Direct evidence for epithelial–mesenchymal transitions in breast cancer. Cancer Res. 2008;68(3):937–945. doi:10.1158/0008-5472.CAN-07-2148

[28] Thiery JP. Epithelial-mesenchymal transitions in tumour progression. Nat Rev Cancer. 2002;2:442–454. doi:10.1038/nrc822

[29] Kroepil F, Fluegen G, Totikov Z, Baldus SE, Vay C, Schauer M, Topp SA, Esch JS, Knoefel WT, Stoecklein NH. Down-regulation of CDH1 is associated with expression of SNAI1 in colorectal adenoma. Plos One 2012;7(9):e46665. doi:10.1371/journal.pone.0046665

[30] Bostanci Z, Mack Jr RP, Lee S, Soybel DI, Kelleher SL. Paradoxical zinc toxicity and oxidative stress in the mammary gland during marginal dietary zinc deficiency. Reprod Toxicol. 2015;54:84–92. doi:10.1016/j.reprotox.2014.07.076

[31] Qin D, Tan L, You Q, Liu X. Expression of heat shock protein 27 and proliferating cell nuclear antigen in human retinoblastoma. Wspolczesna Onkol 2013;17(2):144–149. doi:10.5114/wo.2013.34617

[32] Biaggio VS, Alvarez-Olmedo DG, Perez Chaca MV, Salvetti NR, Valdez SR, Fanelli MA, Ortega HH, Gomez NN, Gimenez MS. Cytoprotective mechanisms in rats lung parenchyma with zinc deprivation. Biometals. 2014;27(2):305–15. doi:10.1007/s10534-014-9713-z

[33] Zhang L, Liu Q, Yuan X, Wang T, Luo S, Lei H, Xia Y. Requirement of heat shock protein 70 for inducible nitric oxide synthase induction. Cell Signal. 2013;25(5):1310–7. doi:10.1016/j.cellsig.2013.02.004

[34] Biaggio VS, Salvetti NR, Pérez Chaca MV, Valdez SR, Ortega HH, Gimenez MS, Gomez NN. Alterations of the extracellular matrix of lung during zinc deficiency. Br J Nutr. 2012;14:108(1):62–70. doi:10.1017/S0007114511005290

[35] Ng IOL, Lai ECS, Fan ST, Ng M, Chan ASY, So MKP. Prognostic significance of proliferating cell nuclear antigen expression in hepatocellular carcinoma. Cancer. 1994;73(9):2268–2274.

[36] Han YH, Gao B, Huang JH, Wang Z, Guo Z, Jie Q, Yang L, Luo ZJ. Expression of CD147, PCNA, VEGF, MMPs and their clinical significance in the giant cell tumor of bones. Int J Clin Exp Pathol. 2015;8(7):8446–8452. eCollection 2015.

[37] Gumulec J, Masarik M, Adam V, Eckschlager T, Provaznik I, Kizek R. Serum and tissue zinc in epithelial malignancies: a meta-analysis. Plos One. 2014;9(6):e99790. doi:10.1371/journal.pone.0099790

[38] Catalani S, De Palma G, Mangili A, Apostoli P. Metallic elements in lung tissues: results of a meta-analysis. Acta Biomed. 2008;79(1):52–63.

[39] Cousins RJ, Liuzzi JP, Lichten LA. Mammalian zinc transport, trafficking, and signals. J Biol Chem. 2006;281(5):24085–24089. doi:10.1074/JBC.R600011200

[40] Franklin RB, Costello LC. The important role of the apoptotic effects of Zinc in the development of cancers. J Cell Biochem. 2009;106(5):750–757. doi:10.1002/jcb.22049

[41] Peters S, Adjei AA, Gridelli C, Reck M, Kerr K, Felip E. Metastatic non-small-cell lung cancer (NSCLC): ESMO Clinical Practice Guidelines for diagnosis, treatment and follow-up. Ann Oncol. 2012;23(7):vii56–vii64. doi:10.1093/annonc/mds226

[42] Kocdor H, Ates H, Aydin S, Cehreli R, Soyarat F, Kemanli P, Harmanci D, Cengiz H Kocdor MA. Zinc supplementation induces apoptosis and enhances antitumor efficacy of docetaxel in non-small-cell lung cancer. Drug Des Dev Ther. 2015;9:3899–3909. doi: 10.2147/DDDT.S87662

[43] Killile AN, Killilea DW. Zinc deficiency reduces paclitaxel efficacy in LNCaP prostate cancer cells. Cancer Lett. 2007;258:25870–25879.

[44] Uzzo RG, Leavis P, Hatch W, Gabai VL, Dulin N, Zvartau N, Kolenko VM. Zinc inhibits nuclear factor-kB activation and sensitizes prostate cancer cells to cytotoxic agents. Clin Cancer Res. 2002;8:3579–3583.

[45] Li G, Zhao J, Peng X, Liang J, Deng X, Yuxiang Chen Y. Radiation/paclitaxel treatment of p53-abnormal non-small cell lung cancer xenograft tumor and associated mechanism. Cancer Biother Radiopharm. 2012;27(4):227–233.

[46] John E, Laskow TC, Buchser WJ, Pitt BR, Basse PH, Butterfield LH, Kalinski P, Lotze MT. Zinc in innate and adaptive tumor immunity. J Transl Med. 2010;8:118. doi: 10.1186/1479-5876-8-118

[47] Schwartz M. Role of trace elements in cancer. Cancer Res. 1975;35(11 Pt. 2):3481–3487.

[48] Margalioth EJ, Schenker JG, Chevion M. Copper and zinc levels in normal and malignant tissues. Cancer Sci. 1983;52:868–872.

[49] Artandi SE, DePinho RA. Telomeres and telomerase in cancer. Carcinogenesis. 2010;31:9–18. doi:10.1093/carcin/bgp268

[50] Meyne J, Ratliff RL, Moyzis RK. Conservation of the human telomere sequence (TTAGGG)n among vertebrates. Proc Natl Acad Sci USA. 1989;86:7049–53. doi:10.1073/PNAS.86.18.7049

[51] Diotti R, Loayza D. Shelterin complex and associated factors at human telomeres. Nucleus. 2011;2:119–35. doi:10.4161/nucl.2.2.15135

[52] Sedivy JM. Can ends justify the means? Telomeres and the mechanisms of replicative senescence and immortalization in mammalian cells. Proc Natl Acad Sci USA. 1998;95(16):9078–9081.

[53] Feng J, Funk WD, Wang SS, Weinrich SL, Avilion AA, Chiu CP, Adams RR, Chang E, Allsopp RC, Yu J, et al. The RNA component of human telomerase. Science. 1995;269(5228):1236–1241. doi:10.1126/science.7544491

[54] Masutomi K, Hahn WC. Telomerase and tumorigenesis. Cancer Lett. 2003;194:163–72.

[55] Vallee BL, Falchuk KH. The biochemical basis of zinc physiology. Physiol Rev. 1993;73:79–118.

[56] Lekili M, Ergen A, Celebi I. Zinc plasma levels in prostatic carcinoma and BPH. Int Urol Nephrol. 1991;23:151–154.

[57] Liang JY, Liu YY, Zou J, Franklin RB, Costello LC, Feng P. Inhibitory effect of zinc on human prostatic carcinoma cell growth. Prostate. 1999;40:200–207.

[58] Nemoto K, KondoY, Himeno S, Suzuki Y, Hara S, Akimoto M, Imura N. Modulation of telomerase activity by zinc in human prostatic and renal cancer cells. Biochem Pharmacol. 2000;59:401–405.

[59] Zarghami N, Hallajzadeh J, Samadzadeh S, Hasanzadeh D, Jabbarzadeh S. Study of correlation between serum zinc levels and telomerase activity in bladder cancer patients. Med J Islam World Acad Sci. 2005;15(2):47–54.

[60] Whelan P, Walker BE, Kelleher J. Zinc, vitamin A and prostatic cancer. Br J Urol. 1983;55:525–528.

[61] Prasad AS, Beck FW, Doerr TD, Shamsa FH, Penny HS, Marks SC, Kaplan J, Kucuk O, Mathog RH. Nutritional and zinc status of head and neck cancer patients: an interpretive review. J Am Coll Nutr. 1998;17:409–418.

[62] Oyama T, Matsuno K, Kawamoto T, Mitsudomi T, Shirakusa T, Kodama Y. Efficiency of serum copper/zinc ratio for differential diagnosis of patients with and without lung cancer. Biol Trace Elem Res. 1994;42:115–127.

[63] Jiang J, Lv W, Ye X, Wang L, Zhang M, Yang H, Okuka M, Zhou C, Zhang X, Liu L, Li J. Zscan4 promotes genomic stability during reprogramming and dramatically improves the quality of iPS cells as demonstrated by tetraploid complementation. Cell Res. 2013;23(1):92–106. doi:10.1038/cr.2012.157

[64] Zalzman M, Falco G, Sharova LV, Nishiyama A, Thomas M, Lee SL, Stagg CA, Hoang HG, Yang HT, Indig FE, Wersto RP, Ko MS. Zscan4 regulates telomere elongation and genomic stability in ES cells. Nature. 2010;464(7290):858–63. doi:10.1038/nature08882

[65] Lee K, Gollahon LS. Zscan4 interacts directly with human Rap1 in cancer cells regardless of telomerase status. Cancer Biol Ther. 2014;15(8):1094–1105. doi:10.4161/cbt.29220

[66] Li K, Zhang J, Ren JJ, Wang Q, Yang KY, Xiong ZJ, Mao YQ, Qi YY, Chen XW, Lan F, Wang XJ, Xiao HY, Lin P, Wei YQ. A novel zinc finger protein Zfp637 behaves as a repressive regulator in myogenic cellular differentiation. J Cell Biochem. 2010;110(2): 352–62. doi:10.1002/jcb.22546

[67] Gao B, Li K, Wei YY, Zhang J, Gao JP, Li YY, Huang LG, Lin P, Wei YQ. Zinc finger protein 637 protects cells against oxidative stress-induced premature senescence by mTERT-mediated telomerase activity and telomere maintenance. Cell Death Dis. 2014;5:e1334. doi:10.1038/cddis.2014.298

Section 2

Iron Deficiency

Chapter 4

Iron Nutrition, Oxidative Stress, and Pathogen Defense

Maria Augusta Naranjo-Arcos and Petra Bauer

Additional information is available at the end of the chapter

http://dx.doi.org/10.5772/63204

Abstract

Adaptation is a challenge that plants have to undergo in order to survive in difficult environments. Nutrient deficiency, stress, and microorganism attack are abiotic and biotic factors that frequently impair plant wellness, which is reflected by low crop yield and quality. Poor crops in turn affect human nutrition. To solve these problems, it is necessary to understand the molecular and physiological mechanisms of nutrient uptake and adaptation to stress. With this knowledge, we may have the possibility to generate new plants, which offer better yield due to their better health. This chapter summarizes and compares iron uptake and assimilation as well as pathogen responses in plants and humans. We also discuss novel approaches for improving crops in the context of human food quality.

Keywords: Iron uptake/absorption, Breeding, ROS, Pathogen

1. Introduction

Organisms have specific ways to assimilate necessary nutrients. Animals, including humans, have to ingest food and process it mechanically and chemically during the digestion. The principal nutrients needed by animals, such as carbohydrates, lipids, proteins, vitamins, and minerals, are found in different sources [1]. A balanced alimentation is hence very important. Lack of essential vitamins or minerals in the diet affects immunity and healthy development. This condition of unproportional alimentation is called undernourishment. Nutrition problems have always been an issue in third world or developing countries. Nowadays, about 104 million children worldwide are underweight (2010). The WHO has a project to reduce by 40% the number of children that are stunted due to the undernourishment until 2025. Currently, in Central Africa over 60% of the population is undernourished, followed by Southeastern

Africa (~40%) and Southern Asia (~20%), which are the regions most affected by undernourishment [2, 3] (**Figure 1**).

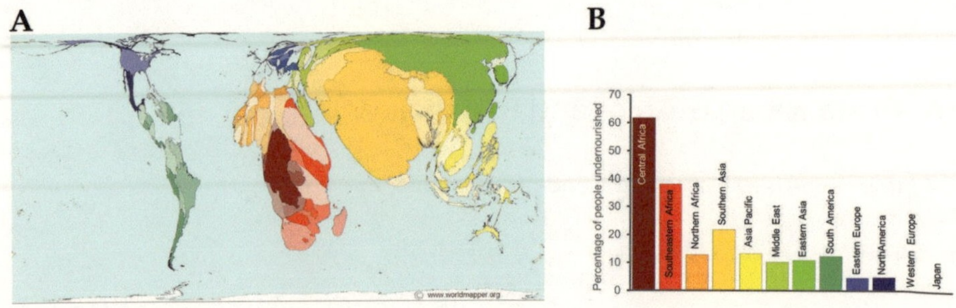

Figure 1. Worldwide map of undernourishment (2000). (A) The map illustrates the resized world corresponding to the number of under-nourished people living in the different regions. (B) The chart shows the percentage of undernourished people living in the different regions. Africa and southern Asia are the most affected regions with almost 50% of all undernourished people worldwide (adapted from [3], Map 178).

In regions with high undernourishment, many people do not have access to a varied diet and their main alimentation consists in only one specific sort of crop. In theory, crops can be fortified by increasing the level of nutrients or uptake-promoting substances in the soil. For example, raising the supply of the essential micronutrient elements Zn, Ni, I, and Se increases their concentration in the grains of several plant products [4]. Unfortunately for other micronutrients, such as Fe, the sole supplementation of the soil with Fe salts is not sufficient to step up the iron quantity in the crops. Foliar fertilization is the best way to increase Fe content in crops, but the cost and effort are not economically interesting [5–7].

Fe deficiency in the form of anemia affects more than 2 billion people on the planet, being the most common nutritional problem in the world (WHO 2005), and regions with Fe deficiency anemia coincide with those of undernourishment, particularly Asia, Africa, and Latin America (compare **Figures 1** and **2**).

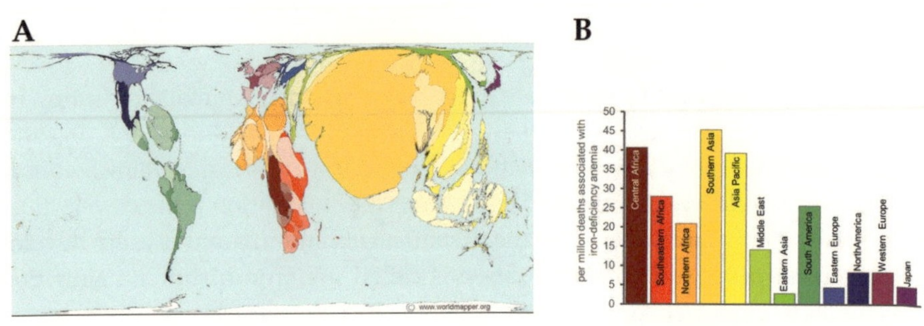

Figure 2. Iron deficiency anemia deaths. (A) The size of the regions corresponds to the number per million deaths associated with Fe deficiency anemia (2002). (B) The chart shows the number of people per million inhabitants living in the different regions. Africa, Asia, and South America are the most affected (adapted from [3], Map 414). (adapted from [3], Map 414).

http://dx.doi.org/10.5772/63204

Due to the poor conditions and the lack of access to diverse nutritious food, it is difficult to counteract this problem by just supplementing the food with iron. Genetic engineering is a suitable approach to fortify plants with organic nutrients. In the case of Fe, it is important to not only find a way to increase the efficiency of the uptake into the plant but also the transport inside the crop, and more importantly to improve the bio-availability of Fe for assimilation in humans [8, 9]. An attractive source and plant crop have to be chosen. For example, cassava or manioc (*Manihot esculenta*) is extensively used by humans for food, livestock, and extraction of starch. Manioc can be cultivated for over 30 years in the same field without fertilizer even in poor soil conditions [10, 11]. Besides, roots can be conveniently stored and also remain in the soil for a long time. Although cassava is one of the basic foods for around 800 million people in the world, it is not a good source for iron and other nutrients [12]. Using genetic engineering, [13] researchers were able to introduce a green algae gene (*FEA1*) in cassava and thus increase the storage of iron in the roots from 10 to 36 ppm. This amount of iron would cover the daily requirements for an adult in a meal of 500 g.

In humans, the iron absorption is strictly regulated. Iron overload is not caused by the consumption of high-iron diets but more by an inadequate ingestion of iron supplements or genetic defects in the regulation of iron homeostasis and iron overload diseases [14]. Difficulties in iron homeostasis produce reactive radicals, better known as reactive oxygen species. Such radicals are capable of damaging almost every molecule in living cells such as DNA, lipids, membrane proteins causing various diseases in humans including vascular diseases and cancer [15].

Iron is an essential element for animals, bacteria, fungi, and plants and similar disease problems as in humans may be found. A competitive situation may arise between organisms when they live in close relationship. This is interesting in host–pathogen interaction systems, where a competition for nutrients between host and pathogen is a determinant for an effective immune system and can affect susceptibility and resistance to a pathogen [16].

2. Iron uptake in plants

Depending on the composition of soil particles, the soil can have different characteristics, some of which define it as fertile. The soil should provide a wide microorganism population, and for most crop plants, a soil pH around 5.5 and 7 is ideal due to the availability of nutrients. The texture of the soil is crucial for the aeration, irrigation, and adequate root proliferation. Its texture is characterized by the amount of sand, silt, and clay particles. High amount of clay particles is necessary to retain essential nutrients and for soil humidity [17]. Around 5.6% of the Earth's crust consists of iron (Fe), belonging to the five most abundant elements. However, the bioavailability of this metal is restricted and plants developed strategies for its mobilization [18]. Iron is a transition metal, and its valence electrons are present in more than one shell so that atoms can be present in several oxidation states [19]. In the nature, iron is present in two biologically important forms, the ferrous (Fe^{2+}) and ferric (Fe^{3+}) form. In acidic environment, iron acts as a reducing factor, whereas in basic medium, it acts as an oxidizing agent [20]. In soil, Fe^{3+} predominates and is attached to silicate structures and hydroxides. In order to take up the Fe ions, plants have strategies to dispatch iron from soil particles, chelate or reduce it and transport it into the plant root cells.

2.1. Strategies of iron uptake in plants

Regarding iron uptake, land plants can be separated into two main groups: Strategy I and Strategy II plants. All plants, except grasses, carry out Strategy I iron uptake. Among them are, for example, tomato (*Lycopersicon esculentum*), tobacco (*Nicotiana tabacum*), potato (*Solanum tuberosum*), and the model organism *Arabidopsis thaliana*. Strategy II plants are all sweet grasses including rice (*Oryza sativa*), maize (*Zea mays*), barley (*Hordeum vulgare*), and wheat (*Triticum* spp.). Studies demonstrated that iron uptake was more efficient in barley (*Hordeam vulgare*) than in cucumber (*Cucumis sativus*) especially at higher pH, which would give Strategy II plants an advantage over Strategy I plants [21].

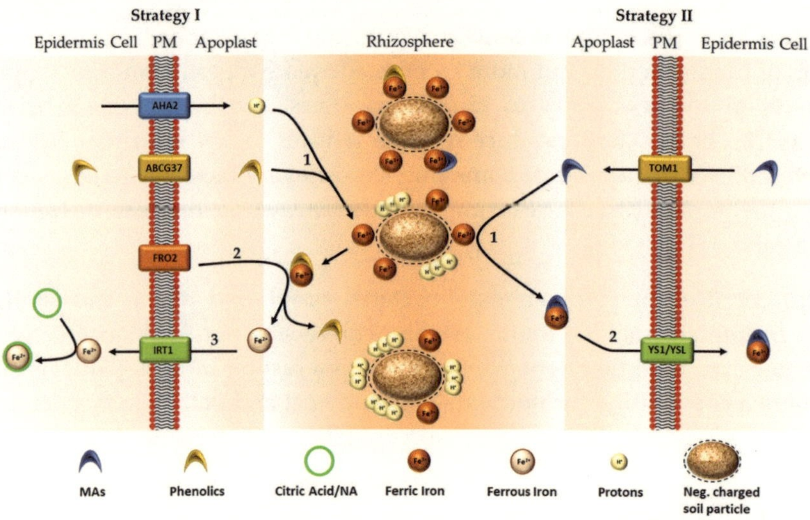

Figure 3. Strategy I and Strategy II iron acquisition in plants. Strategy I plants, exemplified by *Arabidopsis thaliana* (left side), take up iron in three steps: first, in order to liberate Fe^{3+} ions, the proton pump AHA2 acidifies the rhizosphere. The secretion of phenolic compounds through the ABCG37 transporter increases the solubilization of iron. Second, the iron reductase FRO2 reduces Fe^{3+} to Fe^{2+} that finally is transported into the epidermis cell by the iron transporter IRT1. Inside of the plant citric acid or nicotianamine chelate Fe^{3+}/Fe^{2+} for further transport within the plant via xylem or phloem. The iron uptake in strategy II plants, exemplified by *Zea mays* and rice (right side) consists of two steps: firstly, TOM1 exports phytosiderophores into the rhizosphere to solubilize Fe^{3+} ions. The Fe^{3+}/PS complex is transport by the YS1 protein in maize and YSL in other grasses.

In both strategies (**Figure 3**), the proteins required for iron uptake are located in the root epidermis cells [21]. Strategy I plants acidify the rhizosphere by pumping protons, carried out by a proton-ATPase [22]. FERRIC REDUCTASE OXIDASE (FRO2 in *A. thaliana*; LeFRO1 in tomato) is responsible for the reduction of Fe^{3+} to Fe^{2+} which is a crucial step for the iron uptake in Strategy I plants [23–25]. In both strategies, roots enhance iron mobilization by secreting iron-chelating compounds [26]. Among these, many phenolic compounds and flavins are found in Strategy I plants. [27]. The investigation of the effect of phenolic compounds in red

http://dx.doi.org/10.5772/63204

clover (*Trifolium pratense*) showed that the excretion of these molecules is important for the reutilization of apoplastic iron by decreasing the mobilization of iron from roots to shoots [28]. Studies in *A. thaliana*, *Brassica napus*, and *Medicago truncatula* demonstrated that these compounds are related to coumarins such as scopoletin and other derivates as well as flavins. They are produced under iron deficiency conditions, among others, via the action of the feruloyl-CoA 69-hydroxylase1 (F6′H1). Subsequently, the ABC transporter called ABCG37 transports these compounds to the rhizosphere [29–32]. The response of the roots of grasses (Strategy II plants) to iron deficiency is to secrete phytosiderophores (PS) through the phytosiderophore efflux transporter TOM1 [33]. PS are high-affinity iron chelating compounds able to chelate and solubilize ferric iron (Fe^{3+}). The most well-known PS are members of the mugineic acid family (MA) and arvenic acid (AA) [34]. Nicotianamine synthase (NAS) is an important enzyme that catalyzes the fusion of three S-adenosyl methionine molecules (SAM) to form the MA precursor nicotianamine (NA), a non-proteinogenic amino acid [35].

In maize, the first highly specific proton-coupled PS transporter identified was the yellow stripe 1 (YS1). It transports the Fe^{3+}/PS as well as Fe^{3+}/NA complex into the cells [36–38]. Further investigation revealed closely related transporters, yellow stripe 1-like (YSL), in barley and rice [39–41]. The last step of iron uptake in Strategy I plants is the transport of reduced/chelated Fe handled by the IRON-REGULATED TRASNPORTER 1 (IRT1) [42–44].

In contrast to the other Strategy II plants, rice represents a special case because this plant has the ability to take up both Fe^{3+}/PS and Fe^{2+} from the soil. Rice produces lower amounts of PS (2′-deoxymugineic acid DMA) than other grasses, but has two genes encoding for proteins similar to the *Arabidopsis* IRT1, OsIRT1, and OsIRT2. OsIRT proteins were found to be located in the root plasma membrane, and they are able to transport Fe^{2+}. However, rice plants are usually not forced to reduce iron before transport because they grow in submerged conditions where Fe^{2+} is more abundant than Fe^{3+} [45].

Once iron enters the symplast of the epidermal root cells, it diffuses across the plasmodesmata to reach the vascular tissues. The IRON-REGULATED PROTEIN 1 (also known as ferroportin FPN1) IREG1/FPN1 loads Fe into the xylem [46]. The root-specific protein FERRIC REDUCTASE DEFECTIVE 3 (FRD3) mediates the efflux of citrate into the xylem. There, citrate chelates Fe and this complex is transported with the transpiration stream to the upper parts of the plants [47–49]. In order to reach developing organs where the xylem is not yet formed, for example, meristem of young leaves or seeds [50], the Fe is loaded into the phloem and chelated with NA [51–53]. Potential transporters of iron between leaves and sinks are the OLIGOPEPTIDE TRANSPORTER 3 (OPT3) [54] and YSL proteins [55]. In Arabidopsis, NA-chelated Fe may be transported from the phloem to flowers and seeds via the AtYSL1 and AtYSL3 transporter [56, 57], and in rice, this is performed by OsYSL2 [58].

Immediately after reaching the tissue of destination, Fe has to be stored in cell compartments where utilization and storage need to be coordinated. The Fe-transporter FPN2 and VACUOLAR IRON TRANSPORTER 1 (VIT1) [59] are responsible for import of iron into the vacuole, while NATURAL RESISTANCE-ASSOCIATED MACROPHAGE PROTEIN 3 and 4 (NRAMP3 AND NRAMP4) mediate its export [60, 61]. Probably, in the vacuoles, iron is chelated with phytates. Photosynthesis, the electron transport chain and synthesis of chlorophyll require an enormous amount of Fe. Therefore, the majority of iron is supplied to the chloroplasts [62]. The transport of Fe into the chloroplast requires first its reduction mediated by FRO7, followed by its transport performed by the transporter PERMEASE IN CHLOROPLAST 1 (PIC1) [63, 64]. In the chloroplast, Fe is sequestered in ferritin (FER), which is macroprotein complexes able to store up to 4500 iron atoms and present in animals, plants, fungi, and bacteria [65].

2.2. Regulation of the iron uptake in plants

Many transcription factors are responsible for the proper iron homeostasis. The main regulator of the iron uptake is in *A. thaliana* the basic helix-loop-helix (bHLH) protein FER-LIKE IRON DEFICIENCY-INDUCED TRANSCRIPTION FACTOR (AtFIT) [66–68] and in tomato the LeFER [69, 70]. FIT and FER interact with the bHLH proteins of the subgroup Ib and with SlbHLH069, respectively, to activate the transcription of the Fe reductase and the Fe transporter genes in roots [71–73]. The *Arabidopsis* bHLH Ib subgroup transcription factors comprise bHLH038, bHLH039, bHLH100, and bHLH101 [74], which share partial redundant functions in iron homeostasis [75]. There is a large number of genes regulated by FIT and iron deficiency [66, 76, 77]. FIT-dependent genes, partly also under regulation by other iron-regulated transcription factors, are *IRT1, FRO2* [67], *KELCH REPEAT PROTEIN, MTPA2, CYP82C4,* among others [78]. In contrast, the four bHLH subgroup Ib genes are not regulated by FIT. Their high transcript levels in the *fit-3* mutant compared to the wild type are rather due to the iron deficiency [75]. These genes are co-regulated with other known iron homeostasis FIT-independent genes such as *PYE, BTS, FRO3, NRAMP4,* and *NAS4* [66, 79–81].

POPEYE (PYE) and BRUTUS (BTS) are tightly related to iron homeostasis. Both genes are upregulated at –Fe conditions; however, they have opposite functions. PYE acts as a transcription factor positively regulating iron status of the plant, while BTS has repressing effects on the iron homeostasis [82]. BTS belong to the RING E3 ligases proteins which have a hemerythrin group and are able to bind Fe and Zn [83]. Both BTS and PYE interact with IL3, bHLH104, and bHLH115. This interaction might occur, according to the requirements to fine-tune iron homeostasis [82, 84].

Other important regulators of the iron uptake are the redundant MYB10 and MYB72 transcription factors, which are upregulated under low iron conditions. *MYB72* counts as a direct target of FIT and regulates *NAS4* and *NAS2* [81, 85, 86]. Furthermore, MYB72 regulates also the transcription of *BGLU42* that is involved in the production of phenolic compounds, which are excreted by the root to mobilize iron from insoluble sources [30, 81, 87].

3. Iron absorption in humans

In healthy humans, iron represents around 40 mg/kg body weight [88]. Most iron contained in the human body (70%) is circulating with the erythrocytes in form of hemoglobin, around

10% in the muscles as myoglobin, cytochrome, and iron-containing enzymes and the residual 20% as ferritin [89, 90]. The daily-recommended iron dosage for healthy adults is 8 mg for men, 18 mg for women, and 27 mg for pregnant women [91]. Iron absorption efficiency varies depending on the iron type (heme iron or nonheme iron), iron content of the food, iron status of the body, and consumption of iron-absorption inhibitors or enhancers [92]. Meat, poultry, and fish contribute to heme iron while all vegetables, cereals, and legumes with the inorganic oxidized ferric form (Fe^{3+}) (nonheme iron). The bioavailability of heme iron is around fivefold better than nonheme iron, even though the iron content of some plant aliments is much higher than animal food sources [93, 94]. Iron-rich plant aliments are often derived from leafy green vegetables because chloroplasts contain high amounts of metalloproteins that function in the electron transport chain. Seeds or whole grains can also be a good source of iron, which is stored in the form of iron phytate or ferritin in the seed coat or embryo [95].

Many substances consumed simultaneously such as phytic acid and polyphenols impair the bioavailability of nonheme iron [96–98]. Phytic acid (myo-inositol 1,2,3,4,5,6-hexakisphosphate) constitutes the principle phosphorus compound in seeds. Under physiological conditions, these strongly negative compounds form salts with cations such as Ca, Zn, Fe, or Cu to form phytates. In seeds, these salts are found principally in the aleurone layer providing sufficient nutrient sources for the germination [96, 98, 99].

Polyphenols such as tannic and chlorogenic acids are phytochemicals and are mostly present in tea, coffee, red wine, vegetables, fruits, and herbs [100, 101]. These compounds can act as anti-nutrients and inhibit iron absorption into the enterocytes. On the other hand, they have antioxidant properties useful for the human body [102].

Thus, meals containing legumes or whole grains may prevent the proper iron absorption [103]. Therefore, it is necessary to include aliments containing iron-absorption enhancers to a meal. Among these pro-nutrients are plant compounds such as ascorbic acid (vitamin C) and β-carotene (pro vitamin A), but also muscle tissue [104, 105]. Ascorbic acid acts as a reduction factor as well as an iron-chelator facilitating iron transport [106, 107]. It was shown that including β-carotene to the meal improved the iron absorption up to three fold. This effect was observed even if inhibitor-containing food was incorporated [105].

Humans take up heme and nonheme Fe (**Figure 4**). Although absorption of heme iron has not been well described, we know that heme is probably able to cross the lipid bilayer of the cells, but it might also be absorbed via endocytosis as an entire porphyrin structure [108, 109]. Furthermore, a heme carrier protein 1 (HCP1) was described as a mediator for heme transport localized on the enterocytes [110]. However, it was shown one year later that this protein just transports heme incidentally and the actual function is the transport of folate (proton-coupled folate transporter, PCFT). The authors thus named this protein PCFT/HCP1 [111]. Once heme enters the cytoplasm, heme-oxygenases degrade it to release ferrous iron [109, 112], which then binds to ferritin for storage. Since humans do not have an excretion pathway for iron, its excess bound to ferritin is eliminated through the gastrointestinal tract [112].

The iron absorption is triggered when the body-iron sources diminish due to bleeding, inflammation, anemia, or hypoxia. As explained before, nonheme iron from plant food is

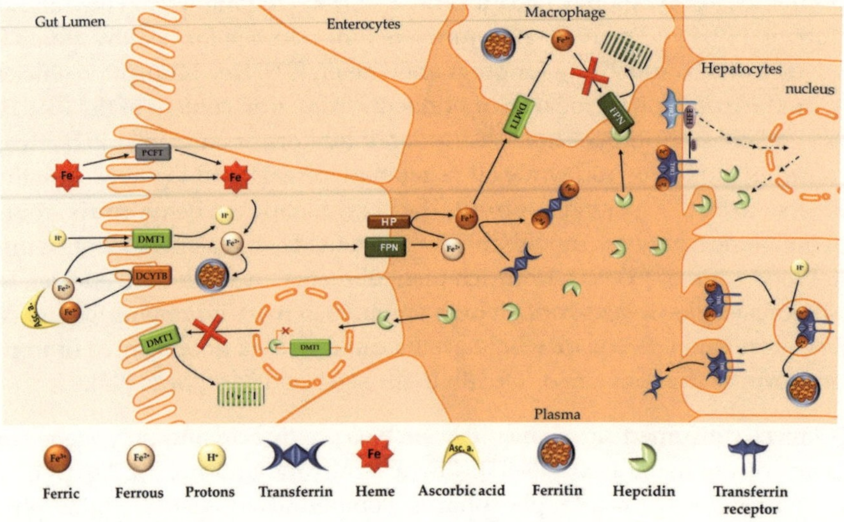

Figure 4. Iron absorption in humans. Fe from the meal exists in form of heme Fe or free Fe^{3+}. Heme is transported via the HCP1/PCFT, and probably absorbed by endocytosis or passing the lipid bilayer. Fe^{3+} from plant sources is first reduced by DCYTB. Ascorbic acid enhanced the reduction of iron in the lumen when taken together with meals. Divalent metal transporters such as DMT1 or NRAMP2 take up the reduced Fe into the cells. Fe is stored in form of ferritin or FPN further transports it into the blood plasma. There, Fe^{2+} is oxidized back to Fe^{3+}, which is either taken up by DMT1 transporter from macrophages or trapped by Tf for the circulation with the plasma. Cells with iron necessity sense and bind the loaded Tf (holotransferrin) with the TfR, which is internalized via endocytosis. Acidification of the vesicles causes the release of Fe, which binds ferritin and the empty TfR is recycled back to the plasma membrane. Iron sufficiency in the cells promotes the complexation of HFE with TfR1 and blocks further binding of holotransferrin. HFE translocate to TfR2 and starts the signal cascade for the production of hepcidin. Hepcidin inhibits the transcription of DMT1 and triggers its internalization and degradation. On the macrophages, hepcidin causes the internalization and degradation of FPN as well.

present as ferric iron (Fe^{3+}). Prior transport, the ingested iron is reduced to ferrous iron (Fe^{2+}) by iron reductase duodenal cytochrome b (DCYTB) which is able to reduce Cu as well [113]. When food rich in vitamin C is included to the meal, the ascorbic acid supports the reduction of iron [107]. The divalent metal transporter 1 (DMT1 or NRAMP2) is responsible for the uptake of Fe^{2+} as well as other divalent ions such as Mn, Co, Cu, Zn, and Cd and protons into the enterocytes [114]. Once in the cytoplasm, Fe^{2+} is bound to ferritin for storage or if, necessary transported to the plasma via basolateral transporter ferroportin (FPN). Hephaestin (HP), a multicopper ferroxidase, oxidizes Fe^{2+} to Fe^{3+} allowing its binding to transferrin (Tf) and the circulation in the plasma [115, 116]. This carrier owns two metal-binding sites found on the N- and C-terminal parts of the molecule. Specific Fe-Tf membrane receptors (TfR) recognize the holotransferrin and the Tf/TfR complex is internalized within the cells via clathrin-coated pits endocytosis. The iron is released from Tf through acidification of the vesicles by a proton pump and can be either stored as ferritin or used to cover the needs of the cell. Finally, the empty Tf/TfR complex moves to the cell surface where the apotransferrin is released to the plasma and charged with new iron [117, 118]. Under levels of iron sufficiency, the protein HFE (hereditary hemochromatosis protein) complexes with the TfR1 blocking the binding of transferrin and iron uptake into the cell. By increased holotransferrin concentration in the plasma, it competes with HFE causing its dissociation. HFE then translocate to TfR2 and "inform" the cell about the elevated Fe-Tf status. This HFE–TfR2 complex starts a signal cascade that regulates the transcription of hepcidin [119–121]. The level of hepcidin in blood modulates the signal for the enterocytes to transport iron into the plasma. Hepcidin is a peptide hormone produced in the liver and negatively regulated during inflammation, anemia, and hypoxia [122, 123]. Under conditions of iron sufficiency, hepatocytes secrete hepcidin to block iron absorption of the enterocytes and transport from macrophages. It was shown that hepcidin inhibits DMT1 transcription and promotes protein internalization followed by its degradation [124, 125]. Besides, hepcidin binds ferroportin (FPN) from macrophages causing its internalization and degradation [126–128]. Hepcidin regulation is still unclear, but there is evidence that the transferrin receptor (TfR), matriptase-2 and hemojuvelin increase the hepcidin expression level [129–131]. Due to the high number of proteins involved in Fe absorption, it is not surprising that several human genetic diseases result in an enhanced Fe absorption, leading to the different types of disorders called hereditary hemochromatosis (HH). The reasons for these disorders are mutations in genes involved in the hepcidin–ferroportin signal transduction such as *HFE*, hepcidin gene *HAMP*, *TfR2*, and ferroportin *FPN*. These mutations disturb the hepcidin-mediated downregulation of ferroportin, the iron transporter responsible for the iron load into the plasma. Furthermore, mutations in genes coding for proteins involved in the iron transport cause insufficient supply of iron for heme synthesis. The consequence is anemia and downregulation of hepcidin despite iron overload [121, 132].

4. Oxidative stress

4.1. Reactive oxygen species (ROS)

The chemical transformation of vital substances in the normal metabolism creates free radicals, which can be any chemical species with one or more unpaired electrons. The free radicals include the hydrogen atom, as the simplest radical, most transition metals and the oxygen molecule. Oxygen radicals, the so-called reactive oxygen species (ROS), are the most common radicals produced in an organism. Among these, the most relevant are the superoxide ($\cdot O_2^-$), hydrogen peroxide (H_2O_2), and hydroxyl radical ($\cdot OH$) [133, 134]. These radicals are produced under normal physiological, pathological, or stress conditions by specific reductase enzymes or in biochemical processes that involve an electron transport chain, for example, photosynthesis, respiration, or oxidative phosphorylation [135]. **Figure 5** shows the complete reduction from oxygen to water with the intermediate radicals. The first step is the reduction of molecular oxygen (O_2) to the relatively stable and reactive superoxide anion (O_2^-). In physiological conditions, O_2^- undergoes spontaneously or enzymatically a dismutation reaction and forms hydrogen peroxide (H_2O_2), which then gives rise to the most reactive oxidant, the hydroxyl radical (($\cdot OH$) [133, 136, 137].

Cells use ROS as messenger molecules for specific responses. For example, H_2O_2 plays an important role in the signal transduction for the immune response of both humans and

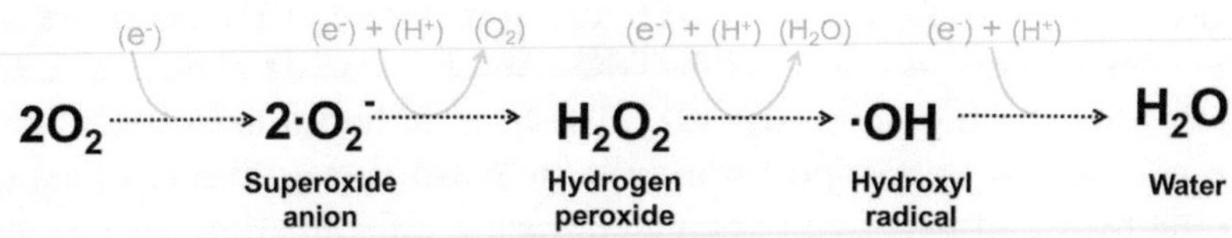

Figure 5. Reactive oxygen species. During processes such as photosynthesis, respiration, and oxidative phosphorylation electrons are transferred to oxygen for the full reduction to water. This process generates very reactive precursors. The first reduction of oxygen generates superoxide anions. Through a dismutation reaction, these anions are converted to form oxygen and hydrogen peroxide. Under the influence of metals, hydrogen peroxide is decomposed and forms the reactive hydroxyl radicals. Dashed arrows and brackets indicate that chemical reactions are not represented in stoichiometrical manner.

plants [138–141]. In humans, it induces the nuclear presence and the DNA-binding of the transcription factor NF-κB, which activates the transcription of genes involved in inflammatory and immune responses [138]. In plants, the immune response is also supported by the production of ROS. These molecules cross-link cell wall proteins to prevent pathogen entrance and induce the production of phytoalexins, which are small molecules that accumulate in the area of infection and prevent growth and spread of the bacteria. Likewise, ROS promote the induction of the hypersensitive response (HR), so that cells undergo programmed cell death to remove nutrient sources for the pathogen. ROS waves are also involved in systemic resistance responses [139, 142].

On the other hand, high production of ROS leads to the so-called oxidative stress. The stimulus of an organism to unaccustomed environments or situations is called stress. The response to such stress might vary from defense and survival to cell death. Toxins, drugs, pollution, and transition metals, iron, in particular, promote the formation of ROS and are thus called prooxidants. When Fe exists abundant in a free state, it is able to catalyze the production of hydroxyl radicals in a two steps reaction called the Haber–Weiss reaction (**Figure 6**). Fe^{3+} first oxidizes the superoxide anion to oxygen and is reduced itself to Fe^{2+}. Then, Fe^{2+} is oxidized in a Fenton reaction, which catalyzes the split of H_2O_2 and the formation of a hydroxyl radical and a hydroxide ion [143–145].

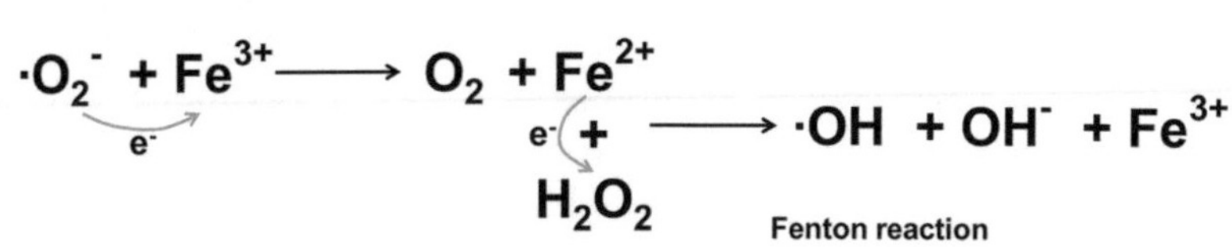

Figure 6. Haber–Weiss reaction. Fe^{3+} catalyzes the production of hydroxyl radicals. Free Fe^{3+} is reduced to $Fe^{2+}$and at the same time oxidizes superoxide anions to oxygen. Fe^{2+} is oxidized and simultaneously cleaves hydrogen peroxide to a hydroxyl radical and a hydroxide ion, which can in turn further react with many molecules.

Ionizing radiation, inflammation, high chemical concentrations, and cellular conditions determine the production of free oxygen radicals. In the case of a chronic inflammation, an overproduction can damage tissues [146].

As mentioned before, Fe uptake in humans and plants is regulated in a way that over-accumulation does not occur in normal cases, except in the iron overload diseases. Regardless, the reason of an iron accumulation, radicals then generate an endless number of diseases such as hepatic cirrhosis, primary liver cancer, diabetes mellitus, arthropathy, cardiomyopy, chronic fatigue, joint pain, impotence, and osteoporosis [147]. In plants, ROS accumulate upon drought, salt stress, cold, heat, heavy metals, high light, ozone, mechanical stress, nutrient deprivation, and pathogen attack. Since plants are not able to escape stresses, they developed mechanisms to adapt and in case of stress-induced ROS to avoid the production of further ROS. In high light conditions, for example, plant diminishes the leaf surface by curling, or upon drought, they close the stomata. If the plant is not able to counteract the production of ROS during the mentioned biotic and abiotic stresses, the cells undergo programmed cell death and the plant may die [148].

4.2. Antioxidants and ROS-scavenging

Low molecular weight molecules and enzymatic proteins are responsible for keeping the balance between harmful and useful ROS. Most ROS are produced in the mitochondria, peroxisomes and, in plants, in chloroplasts as well [148, 149]. Hence, the majority of the ROS-scavenging enzymes are located there [150]. The most important enzymes for ROS-scavenging are the superoxide dismutases (SOD) which catalyze the dismutation of superoxide to form hydrogen peroxide and molecular oxygen. Depending on the compartment, the metal in the active site of this enzyme might change. The MnSOD are present in the mitochondrial matrix, FeSOD in the chloroplast, and CuZnSOD in the rest of the cell [151–153]. The hydrogen peroxide is further processed to water and oxygen by a glutathione peroxidase reaction. Glutathione (GSH) is a tripeptide consisting of glutamic acid, cysteine, and glycine. GSH plays a very important antioxidative role because it is produced in high amount in almost all cells in plants and mammals. It has the capacity to easily oxidize to glutathione disulfide (GSSG). Subsequently, the glutathione reductase converts the GSSG back to GSH. Additionally,

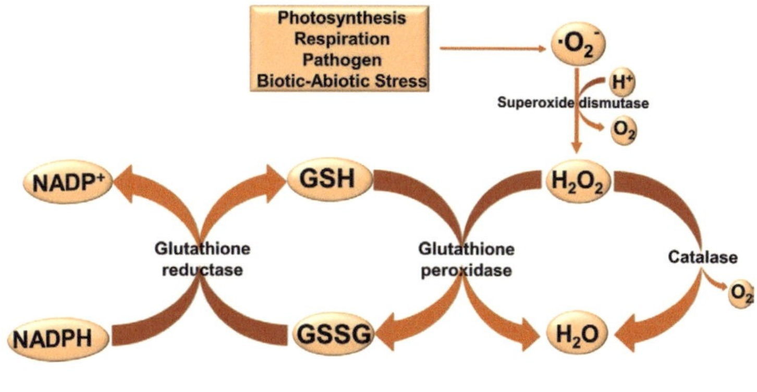

Figure 7. ROS-Scavenging. SOD are responsible for the dismutation of the reactive superoxide to oxygen and hydrogen peroxide. The glutathione reductase reduces $NADP^+$ to NADPH and oxidizes GSH to GSSG. The last served as a substrate for the glutathione peroxidase which decomposes hydrogen peroxide to water and oxygen. Biochemical reactions are not represented in stochiometrical manner.

catalase is a very efficient enzyme to metabolize the hydrogen peroxide to water and oxygen (**Figure** 7) [135, 148, 154, 155].

Another important antioxidant is ascorbic acid (vitamin C) which can react with hydroxyl radicals, superoxide, and oxygen. All animals, except primates and guinea pigs, are able to produce ascorbic acid, and in plants, it accumulates in large amounts. In addition, ascorbic acid reduces prolin and α-tocopherol (Vitamin E) which are important antioxidants as well [156]. In plants, prolin acts as an osmotic agent under salinity and drought. It reacts with hydroxyl radicals to form hydroxyprolin and water, protecting the plant against the radicals [157, 158]. The production of ROS is one mechanism to protect the organism against pathogens attack. The following section will give an overview about the principal mechanisms of pathogen defense in humans and plants.

5. Pathogen defense

The principles of immune responses in plants and mammals differ in several aspects. Mammals have a circulatory system with specialized killing and memory cells as a part of the very effective adaptive immunity. Plants rely on an effective innate immunity.

Plant PRR	Bacterial Antigen	Human PRR
n.d.	Bacterial lipoprotein	TLR 1
LYM1 LYM3 CERK1	Bacterial peptidoglycans	TLR 2
n.d.	Double stranded RNA	TLR 3
LORE	Lipopolysaccharides	TLR 4
FSL2	Bacterial flagella	TLR 5
n.d.	Bacterial lipoprotein	TLR 6
n.d.	Viral single strand RNA	TLR 7/8
n.d.	Bacterial viral DNA	TLR 9
n.d.	n.d.	TLR 10
n.d.	Profilin	TLR 11

n.d. = not determined, PRR = Pattern recognition receptors, TLR = Toll-like receptor. References can be found in the text

Table 1. Bacterial PAMPs and the corresponding PRRs from plants and humans.

The first protection against pathogens is generally to prevent pathogen entrance into the organisms using physical barriers. In plants, callose, lignin, and other phenolics reinforce the cell walls of wood and xylem vessels after pathogen attack or mechanical damage thereby preventing the spread of the pathogen and its toxins [159–162]. There are many barriers, which serve for the first pathogen defense in humans. For example, the skin responds with epithelial peals, drying out, and changes of the pH [163]. Eyes and the respiratory tract produce fluids to wash or catch pathogen invaders [164, 165]. If these barriers fail, the innate immune system comes into place, which is present in all animals and plants.

The first immediate (basal) immune defense in plants and animals is the innate immunity response. Specialized pattern recognition receptors (PRR) recognize microbe molecules, which are called microbe-associated or pathogen-associated molecular patterns (MAMPs/PAMPs) [166, 167]. The different PRRs recognize many different PAMPs including chitin, lipopolysaccharides, peptidoglycans, and flagellin [166–170], and interestingly, similar patterns are recognized by receptors and proteins in plants and animals (**Table 1**).

In plants, the best-characterized receptor is called FLAGELLIN-SENSING 2 (FLS2) and in Arabidopsis was found to recognize the bacterial flagellin epitope (fls22), a 22-amino acid peptide. FLS2 triggers the immune responses, including cell wall fortification and ROS production, and its signaling cascade has been well characterized in the last years [166, 171–

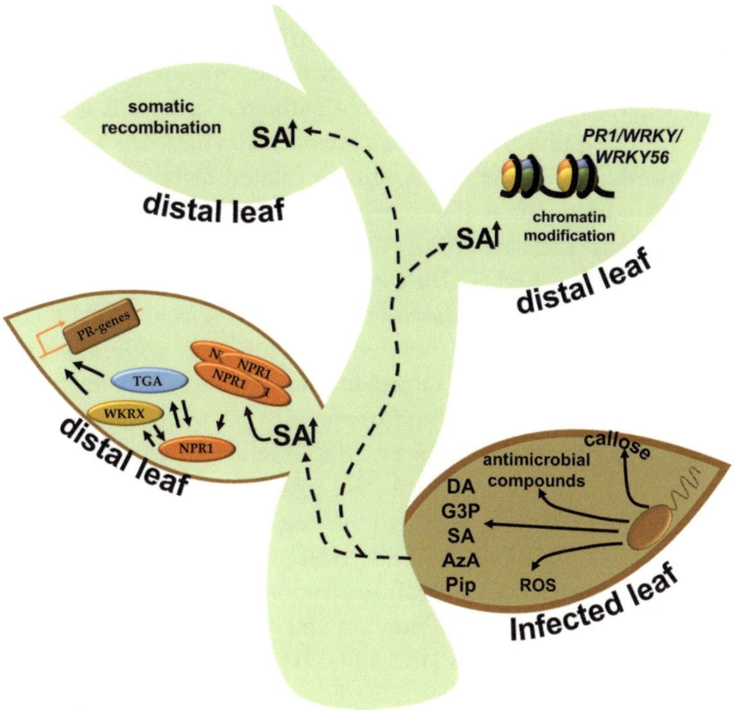

Figure 8. SAR and priming. The attack of a pathogen leads to signals, which give to the plant an immediately protection and a long-lasting immunity. The infected leaf generates callose for the fortification of cell walls, ROS and antimicrobial compounds to kill the bacteria, and defense-related metabolism. These processes are part of the innate immunity. Principally, the metabolite SA is transport through the phloem to the other part of the plants activating the SAR, which is a preventive defense mechanism. This includes the monomerization of NPR1 that in turn activates PR-genes, chromatin modification such as methylation and acetylation of histones and somatic recombination.

173]. The immune response triggered by MAMPs/PAMPs is called PTI (PAMP-triggered immunity). Some bacteria developed a mechanism to bypass the plant pathogen response by injecting effectors (virulence proteins) into plant cells via the bacterial type III secretion system (TTSS). These effectors in turn may be recognized by leucine-rich repeat (NBS-LRR)-type plant proteins and receptors. These resistance factors trigger a second immune response, the effector-triggered immunity (ETI) [174], which may also result in the hypersensitive response (HR), systemic acquired resistance (SAR) including a transcriptional activation of PR proteins, as well as salicylic acid hormone signaling [175–177].

The basic immune output responses are important mechanisms for the surviving of the plant during and after pathogen attack. The HR is an emergency measure, which activates cell death (programmed cell death, PCD) of a limited number of surrounding cells to limit the nutrients and thereby avoid the spread of biotrophic pathogens [178]. Moreover, ETI and PTI activate the so-called systemic acquired resistance (SAR) and the corresponding pathogenesis-related (PR) genes. SAR is a long-distance and long-lasting immune response activated in the entire plant after a local infection [179, 180]. **Figure 8** shows an overview of the forwarding signal of the innate to the acquired immunity.

Fourteen known classes of PR-genes (PR1–PR14) [181] are coordinately expressed to counterattack the pathogens. These genes encode hydrolytic enzymes such as β-1,3-glucanase, chitinase, and plant defensins, which hydrolyze pathogen cell walls and disrupt pathogen membranes [182]. Additionally, the PR-genes encode enzymes needed for the synthesis of callose for the fortification of barriers, and defense-related metabolites such as salicylic acid (SA), diterpenoid dehydroabietinal (DA), a glycerol-3-phosphate (G3P)-dependent factor, azelaic acid (AzA), and pipecolic acid (Pip) [183]. These metabolites are loaded in the phloem and transported to uninfected parts of the plants and "warn" them of an attack [182]. In the distal parts, SA causes the monomerization of the NONEXPRESSOR OF PR GENES 1 (NPR1), allowing its transport to the nucleus. Following, it interacts with TGACG motif-binding protein (TGA) and WRKY transcription factors for the activation of many PR-genes [183–185]. SA accumulation upon pathogen attack and treatment with the protector molecule β-aminobutyric acid (BABA) also provides the plant with a "memory," which allows cells to respond faster and stronger to a secondary challenge. This process is called priming [186, 187]. It was shown that after treatment with a synthetic analog of SA transcript and protein of the MITOGEN-ACTIVATED PROTEIN KINASES 3 AND 6 (MPK3, MPK6) accumulated in upper leaves after infection of lower leaves [188]. It was reported that priming after a certain attack can be mediated through somatic recombination and modification of the chromatin such as acetylation and methylation of histones [182, 189–191] (**Figure 8**).

Humans have an innate immune system as well, which is the first response to a pathogen attack. The innate immune system provides the organism with a rapid but not specific response. Surface barriers like skin, fluids and antimicrobial peptides prevent the entrance of the pathogens into the body [163]. When pathogens pass these barriers, the body attempts to eliminate the source of injury and damaged tissue. This is carried out by the generation of an inflammation, which includes high blood irrigation, immune cells, and mediators. Inflammation is the first immune reaction to a pathogen invasion or injury and corresponds to the innate

http://dx.doi.org/10.5772/63204

immune response, which works in a similar way as in plants [192]. Bacterial components, best known as PAMPs (**Table 1**), are recognized by PRR. They are called in humans toll-like receptors (TLR1-11). These receptors are on the surface of mast cells, macrophages, and dendritic cells [193, 194]. Once a mast cell recognizes a PAMP, it secretes mediating factors, such as histamine or tumor necrosis factor (TNFα). These factors dilate the blood vessels enabling neutrophils to enter the damaged tissue. Bacteria-degrading substances are secreted, and bacteria are removed by phagocytosis [194]. The activated macrophages phagocytose bacteria produce cytokines and prostaglandins, which attract neutrophils, monocytes, and dendritic cells. Cytokines are responsible and induce a rise of the body temperature (fever) because the growth of many pathogens is then compromised. Dendritic cells identify PAMPs and mature to antigen presenting cells (APC). They process the antigens and recruit them to the T-lymphocytes to start the adaptive immune answer. Hence, the dendritic cells are the link between the innate and the adaptive immune system [195].

5.1. Iron and pathogen defense

Pathogens gain from the host all required nutrients. Iron regulation is very important for pathogen survival during infection in plants. Iron plays dual roles for host and pathogen, either as nutrient or as essential cofactor constituent to initiate or avoid immune responses.

The genes *FER2* and *FER1* of a maize fungi *Ustilago maydis*, which encode a high-affinity iron permease and an iron multicopper oxidase respectively, are involved in the iron uptake in this microorganism. Deletion of these genes showed that the infection rate of the fungi in maize plants was impaired, concluding that these genes are crucial for its virulence as well [196]. Furthermore, NPS6 is a virulent gene conserved in many filamentous fungi, which is involved in siderophore biosynthesis and in tolerance to H_2O_2 [197].The requirement to sequestrate iron via siderophore production might be not only to take up iron but also to protect from reactive oxygen species [198].

In the host cells, iron exists mostly in complex with ferritin, transferrin, hemoglobin, and other proteins. It was shown that after a pathogen infection in plants ferritin accumulates, possibly to protect from ROS but also to deprive invaders from iron [199]. During a pathogen attack, iron accumulates in the apoplast to elevate the oxidative response, which in turn activates the expression of PR-genes. The translocation of iron to the apoplast causes intracellular iron deficiency activating iron uptake genes and PR-genes [200]. In mammals, during an infection or inflammation the hepcidin level rises, this stops the iron absorption and stimulates the transport of circulating iron (or free heme from damaged tissue) into the macrophages. This is a mechanism controlled by cytokines [201]. However, many pathogens developed the property to gain iron from transferrin and heme. For example, *V. cholerae* is able to induce the release of heme from hemoglobin. *Hemophilus influenzae* or *Trichomonas vaginalis* trigger release of iron from transferrin [202]. Bacteria stimulate siderophore production, which may serve to solubilize iron inside the host and transport it into the microbial cells [16, 203]. The upregulation of ferritin may help to withstand siderophore action [199].

However, many bacteria can also act in a positive way on plants. In the plant rhizosphere, many beneficial nonpathogenic rhizobacteria are present, which protect the plants against pathogenic microorganisms by secreting antimicrobial components [204]. The content of plant growth-promoting rhizobacteria (PGPR), soil type, and strategy of iron uptake plays a fundamental role for plants in the iron nutrition [205, 206]. In turn, the iron status of the plant influences the rhizobacteria community as well. Studies in barley and tomato showed that the bacterial community in rhizosphere of iron deficient plants is much smaller than of plants grown in iron sufficient conditions. Additional, the different iron uptake strategies of these two plants leads to a different qualitative and quantitative patterns of the rhizobacterial population [206].

Plants may profit from siderophore production of the rhizobacteria for the activation of the induced systemic resistance (ISR), a SA-independent immunological pathogen response in plants [207, 208]. Iron deficiency and ISR are closely related. Transcriptome analysis in *A. thaliana* showed that a high number of genes upregulated upon iron deficiency conditions are also upregulated in plants treated with beneficial bacteria [209]. Similarly, treatment with synthetic siderophores upregulated many gene encoding for WRKY transcription factors and genes required for the iron uptake, such as ferritin, the iron transporter (IRT1), the iron reductase (FRO2), and the NICOTIANAMINE SYNTHASE (NAS) [199, 210]. An overlap between these two processes is represented by the transcription factor MYB72. This transcription factor activates downstream an important component necessary for ISR called BGLU42 (β-glucosidase). BGLU42 is involved in the production and secretion of phenolic compounds from the root to the rhizosphere and is a key component for the activation of the ISR [87]. *MYB72* is strongly upregulated under iron deficiency conditions, and its regulation occurs in a FIT-depended manner [85, 86]. Thus, mutants lacking MYB72, and its close homologue MYB10, were not able to survive in iron deficiency conditions. Moreover, it was shown that MYB72 induces the expression of *NAS4* [51, 85] and *BHLH039* [87].

The effect of siderophores on Fe homeostasis is comparable with treatments with BABA. BABA (β-aminobutyric acid) is a nonprotein amino acid priming the plants against a broad-spectrum of pathogens as well as abiotic stresses such as salt stress and drought. The mechanism of protection is based on the activation of the SA-dependent pathogen response, abscisic acid (ABA)-dependent formation of callose [211–213]. Moreover, BABA affects the Fe-homeostasis upon reducing ferritin and increasing *NAS4* transcript and protein. Plants treated with BABA show similar phenotype, transcription of *IRT1* and *FRO2* and metabolite composition as iron deficient plants. BABA is able to chelate iron and mimic the effect of the pathogen siderophores intensifying the theory that iron scavenging is a strategy for the activation of pathogen responses [210, 214]

An additional overlap between Fe homeostasis and pathogen defense is provided by the plant hormones SA. As mentioned before, SA is necessary for the activation of PR-genes and the priming processes for the long-lasting defense [182, 183, 187]. The transcription factor OBP3 (OBF-BINDING PROTEIN 3) is induced by SA and is involved in plant growth and development [215, 216]. Studies in OBP3-overexpression lines had shown that *BHLH038* and *BHLH039* were strongly upregulated [217]. However, an effect of SA on iron deficiency response regulation has later not been found in the wild type or in the triple mutant *bhlh039 bhlh100 bhlh101* [78].

Figure 9 shows the relationship of Fe homeostasis with the production of ROS and the pathogen response. Fe is required as a cofactor for the production of ROS. The local iron deficiency induced by bacterial siderophores upregulates *MYB72*, which in turn activates the expression of *BHLH039, NAS4,* and *BGLU42*. Additional observations have also led to the conclusion that pathogens and the hosts may compete for Fe and that Fe may be required, at the same time, for inducing defense but also to sustain pathogen infection [207, 208].

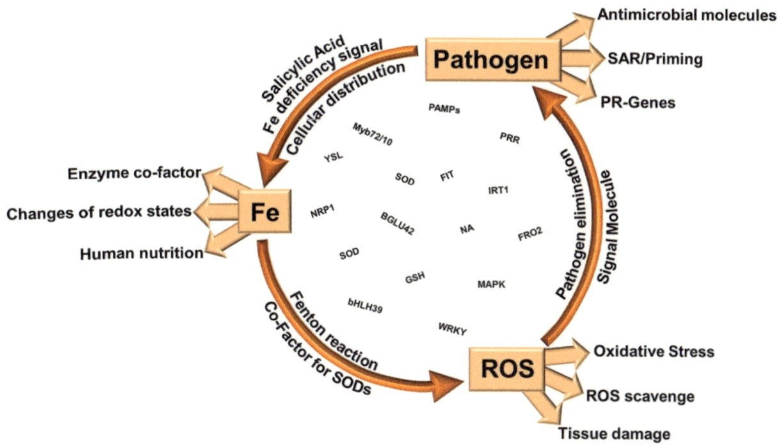

Figure 9. Correlation between Fe, ROS, and pathogen defense. All three processes are tightly related. Besides the initiation of pathogen responses during microorganism attack, Fe translocates to the apoplast causing a local Fe deficiency and increasing Fe uptake from the roots. High levels of Fe lead to the accumulation of ROS, but simultaneously it serves as a co-factor for enzymes involved in the ROS scavenging. ROS molecules support the elimination of pathogens and activate local PR genes. In the center of the ring are represented the genes involved in the three processes.

6. Breeding for better crops

Selection of better plants, genetic engineering, and breeding can also be used for the generation of nutrient-rich crops. A good example is provided by the rice IR68144 line, which presents tolerance against diseases, high yields, and high Fe and Zn content [218]. Studies in Philippine women showed that the consumption of this rice versus the normal diet increased the iron content in the body [219]. Nowadays, the world population grows with great rapidity and the use of common breeding methods may not be sufficient to cover the necessity of better crops and yield. It is possible to increase the iron content of edible organs of plants by improving iron mobilization in the soil and plant, the storage and remobilization in the leaves, grains or fruits.

In rice, the insertion of phytosiderophore synthesis genes from barley increased considerably its tolerance to calcareous soil and Fe and Zn content in the grain [220]. The introduction of a yeast Fe reductase into rice increased likewise yield and the tolerance to a Fe deficiency environment. However, the iron content of the grain was not improved [221]. The *A. thaliana*

IRT1 transports not only Fe but also other divalent metal cations such as Zn, Mn, and Co [222]. Increased activity of Fe transporters may lead partially to an increased iron content in the grain but simultaneously may also compromise plant health due to the accumulation of other metals, which in turn leads to a high ROS production [223]. Furthermore, it is important that in addition to an increased iron content in the plants, attention should also be paid to iron bioavailability for humans.

Seeds have a high amount of phytic acids, which form a complex with iron and inhibit its absorption in the intestinal lumen [97]. Maize plants co-expressing the phytic acid-degrading enzyme phytase from *Aspergillus niger* and ferritin from soybean under the control of an endosperm promoter, increased up to threefold the iron bioavailability in *in vitro* experiment with Caco-2 cells [224]. Similar in rice, the expression of ferritin from *Phaseolus vulgaris,* phytase from *Aspergillus fumigates,* and the endogenous cysteine-rich metallothionein-like protein improved iron bioavailability for humans [225]. Genetic approaches in *Lactuca sativa* (common Lettuce) showed that introducing the soybean ferritin gene is sufficient to increase the iron content in the leaves, boost the growth rate, and enhance the size up to 40% [226].

The pathogen defense of plants can be also improved using genetic methods. PR-genes involved in the production of antimicrobial substances such as chitinases, β-1,3-glucanase, defensins, osmotin, and phytoalexins can be used to increase the strength of crops against biotic stresses [227]. Osmotin is a PR-gene from tobacco expressed under biotic and abiotic stresses, which causes holes in the plasma membrane of several fungi [228, 229]. Overexpression of osmotin in potato leads to a delay of the disease caused by *Hytophthora infestans* [230]. Defensins are cysteine-rich peptide with antimicrobial properties in humans and plants [231, 232]. When expressed in potato, the antifungal protein (AFP) from *Medicago sativa* seeds leads to strong resistance against the fungal pathogen *Verticillium dahlia* [233]. The fungal cell wall is a very complex structure composed of many different proteins as well as chitin and glycan, which are essential components [234]. Chitinases and glucanases are potent antifungal enzymes and hydrolyze fungal cell wall [227]. Apple trees transformed with the endochinitase from *Trichoderma harzianum* present an important reduction in the number of lesions and its area after *Venturia inaequalis* inoculation [235]. β-1,3-glucanase inhibits the grow of many fungi as in the case of flax (*Linum usitatissimum* L.). Several transgenic lines caring the potato β-1,3-glucanase present high resistance against *Fusarium* species, demonstrating that degradation of fungal cell walls is a weapon against fungi [236].

Many abiotic factors are causing oxidative stress in plants. Investigation on the aluminum-induced genes showed that Arabidopsis plants of the ecotype Ler-0 expressing the blue-copper-binding protein (AtBCB), the peroxidase gene (AtPox), or the tobacco glutathione S-transferase gene (parB) were more resistant to treatment with oxidative stress inducing diamide [237]. The overexpression of CuZn-Superoxid dismutase (CuZnSOD), Mn-Superoxid dismutase (MnSOD), and ascorbate peroxidase (APX) in transgenic tobacco plants were more tolerant to the viologen (MV, paraquat)-mediated oxidative damage [238]. Ascorbic acid is an important antioxidant [156] in plants. Tomato (*Solanum lycopersicum*) plants over-expressing GDP-Mannose 3′,5′-epimerase (*SlGME1* and *SlGME2*) are able to accumulate

http://dx.doi.org/10.5772/63204

ascorbic acid and improved the tolerance to viologen stress, cold stress, and better biomass under salt stress.

To date, there are only a few examples of how the genetic engineering is able to improve plant growth and crop yield during situations of pathogen attack and different abiotic stresses such as nutrient deficiency or environmental stresses.

Author details

Maria Augusta Naranjo-Arcos[1] and Petra Bauer[1,2*]

*Address all correspondence to: petra.bauer@hhu.de

1 Institute of Botany, Heinrich-Heine University, Universitätstrasse 1, Düsseldorf, Germany

2 Cluster of Excellence on Plant Sciences (CEPLAS), Heinrich-Heine University, Düsseldorf, Germany

References

[1] Gropper SS, Smith JL. Advanced nutrition and human metabolism. Cengage Learning. 2013;6th Edition(2).

[2] Dorling D, Barford A, Newman M. WORLDMAPPER: the world as you've never seen it before. Visualization and Computer Graphics, IEEE Transactions on. 2006;12(5):757–64.

[3] Dorling D, Newman M, Allsopp G, Barford A, Wheeler B, Pritchard J, et al. Worldmapper.org 2006. Available from: http://www.worldmapper.org.

[4] Bouis HE, Welch RM. Biofortification – a sustainable agricultural strategy for reducing micronutrient malnutrition in the global south. Crop Science. 2010;50(Supplement 1):S-20–S-32.

[5] Mortvedt JJ. Correcting iron deficiency in annual and perennial plants: present technologies and future prospects. Plant and Soil. 1991;130:273–9.

[6] Marschner H, Römheld V. Strategies of plants for acquisition of iron. Plant and Soil. 1994;165:261–74.

[7] Frossard E, Bucher M, Mächler F, Mozafar A, Hurrell R. Potential for increasing the content and bioavailability of Fe, Zn and Ca in plants for human nutrition. Journal of the Science of Food and Agriculture. 2000;80:861–79.

[8] Perez-Massot E, Banakar R, Gomez-Galera S, Zorrilla-Lopez U, Sanahuja G, Arjo G, et al. The contribution of transgenic plants to better health through improved nutrition: opportunities and constraints. Genes & Nutrition. 2013;8(1):29–41.

[9] Zhu C, Naqvi S, Gomez-Galera S, Pelacho AM, Capell T, Christou P. Transgenic strategies for the nutritional enhancement of plants. Trends in Plant Science. 2007;12(12):548–55.

[10] Tracy SM, Cassava, US Department of Agriculture. Washington, D.C. 1903, 167, p. 1–32.

[11] Nweke F, Steven H, Zulu B. Recent growth in African cassava, building on successes in African Agriculture. International Food Policy Research Institute (IFPRI). 2004;Focus 12(Brief3).

[12] Sautter C, Poletti S, Zhang P, Gruissem W. Biofortification of essential nutritional compounds and trace elements in rice and cassava. Proceedings of the Nutrition Society. 2007;65(02):153–9.

[13] Ihemere UE, Narayanan NN, Sayre RT. Iron biofortification and homeostasis in transgenic cassava roots expressing the algal iron assimilatory gene, FEA1. Frontiers in Plant Science. 2012;3:171.

[14] Heath A-L. Health implications of iron overload. Nutrition Reviews. 2003;61(2):45–62.

[15] Halliwell B. Free radicals, reactive oxygen species and human disease: a critical evaluation with special reference to atherosclerosis. British Journal of Experimental Pathology. 1989;70(6):737–57.

[16] Payne SM. Iron acquisition in microbial pathogenesis. Trends in Microbiology. 1993;1(2):66–9.

[17] Jones JB. Plant Nutrition and Soil Fertility Manual . 2nd ed. Boca Raton, FL, USA. 2012. p. 9781439816103.

[18] Guerinot ML, Yi Y. Iron: nutritious, noxious, and not readily available. Plant Physiology. 1994;104(3):815–20.

[19] McNaught AD, Wilkinson A. IUPAC. Compendium of Chemical Terminology, 2nd ed. (the "Gold Book"). Blackwell Scientific Publications, Oxford. 1997.

[20] Pauling L. General Chemistry, 3rd ed. Dover Publication, INC, New York. 1988;3:578–89; 678–93.

[21] Marschner H, Romheld V, Kissel M. Different strategies in higher plants in mobilization and uptake of iron. Journal of Plant Nutrition. 1986;9(3):695–713.

[22] Santi S, Schmidt W. Dissecting iron deficiency-induced proton extrusion in Arabidopsis roots. New Phytol. 2009;183(4):1072–84.

[23] Robinson NJ, Procter CM, Connolly EL, Guerinot ML. A ferric-chelate reductase for iron uptake from soils. Nature. 1999;397(6721):694–7.

[24] Connolly EL, Campbell NH, Grotz N, Prichard CL, Guerinot ML. Overexpression of the FRO2 ferric chelate reductase confers tolerance to growth on low iron and uncovers posttranscriptional control. Plant Physiology. 2003;133(3):1102–10.

[25] Li L, Cheng X, Ling H-Q. Isolation and characterization of Fe (III)-chelate reductase gene LeFRO1 in tomato. Plant Molecular Biology. 2004;54(1):125–36.

[26] RSmheld V, Marschner H. Mechanism of iron uptake by peanut plants. Plant Physiology. 1983;71:949–54.

[27] Yoshino M, Murakami K. Interaction of iron with polyphenolic compounds: application to antioxidant characterization. Analytical Biochemistry. 1998;257(1):40–4.

[28] Jin CW, You GY, He YF, Tang C, Wu P, Zheng SJ. Iron deficiency-induced secretion of phenolics facilitates the reutilization of root apoplastic iron in red clover. Plant Physiology. 2007;144(1):278–85.

[29] Mladenka P, Macakova K, Zatloukalova L, Rehakova Z, Singh BK, Prasad AK, et al. In vitro interactions of coumarins with iron. Biochimie. 2010;92(9):1108–14.

[30] Rodriguez-Celma J, Pan IC, Li W, Lan P, Buckhout TJ, Schmidt W. The transcriptional response of Arabidopsis leaves to Fe deficiency. Frontiers in Plant Science. 2013;4:276.

[31] Fourcroy P, Siso-Terraza P, Sudre D, Saviron M, Reyt G, Gaymard F, et al. Involvement of the ABCG37 transporter in secretion of scopoletin and derivatives by Arabidopsis roots in response to iron deficiency. New Phytologist. 2014;201(1):155–67.

[32] Schmid NB, Giehl RF, Doll S, Mock HP, Strehmel N, Scheel D, et al. Feruloyl-CoA 6'-Hydroxylase1-dependent coumarins mediate iron acquisition from alkaline substrates in Arabidopsis. Plant Physiology. 2014;164(1):160–72.

[33] Nozoye T, Nagasaka S, Kobayashi T, Takahashi M, Sato Y, Sato Y, et al. Phytosiderophore efflux transporters are crucial for iron acquisition in graminaceous plants. Journal of Biological Chemistry. 2011;286(7):5446–54.

[34] Römheld V. Different strategies for iron acquisition in higher plants. Physiologia Plantarum. 1987;70(2):231–4.

[35] Higuchi K, Nakanishi H, Suzuki K, Nishizawa NK, Mori S. Presence of nicotianamine synthase isozymes and their homologues in the root of graminaceous plants. Soil Science and Plant Nutrition. 1999;45(3):681–91.

[36] Curie C, Panaviene Z, Loulergue C, Dellaporta SL, Briat JF, Walker EL. Maize yellow stripe1 encodes a membrane protein directly involved in Fe(III) uptake. Nature. 2001;409(6818):346–9.

[37] Roberts LA, Pierson AJ, Panaviene Z, Walker EL. Yellow stripe1. Expanded roles for the maize iron-phytosiderophore transporter. Plant Physiology. 2004;135(1):112–20.

[38] Schaaf G, Ludewig U, Erenoglu BE, Mori S, Kitahara T, von Wiren N. ZmYS1 functions as a proton-coupled symporter for phytosiderophore- and nicotianamine-chelated metals. Journal of Biological Chemistry. 2004;279(10):9091–6.

[39] Murata Y, Ma JF, Yamaji N, Ueno D, Nomoto K, Iwashita T. A specific transporter for iron(III)-phytosiderophore in barley roots. The Plant Journal. 2006;46(4):563–72.

[40] Inoue H, Kobayashi T, Nozoye T, Takahashi M, Kakei Y, Suzuki K, et al. Rice OsYSL15 is an iron-regulated iron(III)-deoxymugineic acid transporter expressed in the roots and is essential for iron uptake in early growth of the seedlings. Journal of Biological Chemistry. 2009;284(6):3470–9.

[41] Araki R, Murata J, Murata Y. A novel barley yellow stripe 1-like transporter (HvYSL2) localized to the root endodermis transports metal-phytosiderophore complexes. Plant Cell Physiology. 2011;52(11):1931–40.

[42] Eide D, Broderius M, Fett J, Guerinot ML. A novel iron-regulated metal transporter from plants identified by functional expression in yeast. Proceedings of the National Academy of Sciences USA. 1996;93(11):5624–8.

[43] Cohen CK, Fox TC, Garvin DF, Kochian LV. The role of iron-deficiency stress responses in stimulating heavy-metal transport in plants. Plant Physiology. 1998;116(3):1063–72.

[44] Eckhardt U, Marques AM, Buckhout TJ. Two iron-regulated cation transporters from tomato complement metal uptake-deficient yeast mutants. Plant Molecular Biology. 2001;45(4):437–48.

[45] Ishimaru Y, Suzuki M, Tsukamoto T, Suzuki K, Nakazono M, Kobayashi T, et al. Rice plants take up iron as an Fe^{3+}-phytosiderophore and as Fe^{2+}. The Plant Journal. 2006;45(3):335–46.

[46] Morrissey J, Baxter IR, Lee J, Li L, Lahner B, Grotz N, et al. The ferroportin metal efflux proteins function in iron and cobalt homeostasis in Arabidopsis. The Plant Cell. 2009;21(10):3326–38.

[47] Green LS, Rogers EE. FRD3 controls iron localization in Arabidopsis. Plant Physiology. 2004;136(1):2523–31.

[48] Durrett TP, Gassmann W, Rogers EE. The FRD3-mediated efflux of citrate into the root vasculature is necessary for efficient iron translocation. Plant Physiology. 2007;144(1):197–205.

[49] Rogers EE. FRD3, a member of the multidrug and toxin efflux family, controls iron deficiency responses in arabidopsis. The Plant Cell Online. 2002;14(8):1787–99.

[50] Grusak MA. Iron transport to developing ovules of *Pisum sativum* (I. seed import characteristics and phloem iron-loading capacity of source regions). Plant Physiology. 1994;104(2):649–55.

[51] von Wirén N, Klair S, Bansal S, Briat J-F, Khodr H, Shioiri T, et al. Nicotianamine chelates both FeIII and FeII. Implications for metal transport in plants. Plant Physiology. 1999;119(3):1107–14.

[52] Curie C, Cassin G, Couch D, Divol F, Higuchi K, Le Jean M, et al. Metal movement within the plant: contribution of nicotianamine and yellow stripe 1-like transporters. Annals of Botany. 2009;103(1):1–11.

[53] Schuler M, Rellan-Alvarez R, Fink-Straube C, Abadia J, Bauer P. Nicotianamine functions in the Phloem-based transport of iron to sink organs, in pollen development and pollen tube growth in Arabidopsis. The Plant Cell. 2012;24(6):2380–400.

[54] Stacey MG, Patel A, McClain WE, Mathieu M, Remley M, Rogers EE, et al. The Arabidopsis AtOPT3 protein functions in metal homeostasis and movement of iron to developing seeds. Plant Physiology. 2008;146(2):589–601.

[55] Chu HH, Chiecko J, Punshon T, Lanzirotti A, Lahner B, Salt DE, et al. Successful reproduction requires the function of Arabidopsis Yellow Stripe-Like1 and Yellow Stripe-Like3 metal-nicotianamine transporters in both vegetative and reproductive structures. Plant Physiology. 2010;154(1):197–210.

[56] Le Jean M, Schikora A, Mari S, Briat JF, Curie C. A loss-of-function mutation in AtYSL1 reveals its role in iron and nicotianamine seed loading. The Plant Journal. 2005;44(5): 769–82.

[57] Waters BM, Chu HH, Didonato RJ, Roberts LA, Eisley RB, Lahner B, et al. Mutations in Arabidopsis yellow stripe-like1 and yellow stripe-like3 reveal their roles in metal ion homeostasis and loading of metal ions in seeds. Plant Physiology. 2006;141(4):1446–58.

[58] Ishimaru Y, Masuda H, Bashir K, Inoue H, Tsukamoto T, Takahashi M, et al. Rice metal-nicotianamine transporter, OsYSL2, is required for the long-distance transport of iron and manganese. The Plant Journal. 2010;62(3):379–90.

[59] Kim SA, Punshon T, Lanzirotti A, Li L, Alonso JM, Ecker JR, et al. Localization of iron in Arabidopsis seed requires the vacuolar membrane transporter VIT1. Science (New York, NY). 2006;314(5803):1295–8.

[60] Thomine S, Lelièvre F, Debarbieux E, Schroeder JI, Barbier-Brygoo H. AtNRAMP3, a multispecific vacuolar metal transporter involved in plant responses to iron deficiency. The Plant Journal. 2003;34(5):685–95.

[61] Lanquar V, Lelievre F, Bolte S, Hames C, Alcon C, Neumann D, et al. Mobilization of vacuolar iron by AtNRAMP3 and AtNRAMP4 is essential for seed germination on low iron. The EMBO Journal. 2005;24(23):4041–51.

[62] Kim SA, Guerinot ML. Mining iron: iron uptake and transport in plants. FEBS Letters. 2007;581(12):2273–80.

[63] Duy D, Wanner G, Meda AR, von Wiren N, Soll J, Philippar K. PIC1, an ancient permease in Arabidopsis chloroplasts, mediates iron transport. The Plant Cell. 2007;19(3):986–1006.

[64] Duy D, Stube R, Wanner G, Philippar K. The chloroplast permease PIC1 regulates plant growth and development by directing homeostasis and transport of iron. Plant Physiology. 2011;155(4):1709–22.

[65] Briat J-F, Lobréaux S. Iron transport and storage in plants. Trends in Plant Science. 1997;2(5):187–93.

[66] Colangelo EP, Guerinot ML. The essential basic helix-loop-helix protein FIT1 is required for the iron deficiency response. The Plant Cell. 2004;16(12):3400–12.

[67] Jakoby M, Wang HY, Reidt W, Weisshaar B, Bauer P. FRU (BHLH029) is required for induction of iron mobilization genes in *Arabidopsis thaliana*. FEBS Letters. 2004;577(3): 528–34.

[68] Bauer P, Ling HQ, Guerinot ML. FIT, the FER-like iron deficiency induced transcription factor in Arabidopsis. Plant Physiology and Biochemistry. 2007;45(5):260–1.

[69] Ling HQ, Bauer P, Bereczky Z, Keller B, Ganal M. The tomato fer gene encoding a bHLH protein controls iron-uptake responses in roots. Proceedings of the National Academy of Sciences USA. 2002;99(21):13938–43.

[70] Brumbarova T, Bauer P. Iron-mediated control of the basic helix-loop-helix protein FER, a regulator of iron uptake in tomato. Plant Physiology. 2005;137(3):1018–26.

[71] Yuan Y, Wu H, Wang N, Li J, Zhao W, Du J, et al. FIT interacts with AtbHLH38 and AtbHLH39 in regulating iron uptake gene expression for iron homeostasis in Arabidopsis. Cell Research. 2008;18(3):385–97.

[72] Wang N, Cui Y, Liu Y, Fan H, Du J, Huang Z, et al. Requirement and functional redundancy of Ib subgroup bHLH proteins for iron deficiency responses and uptake in *Arabidopsis thaliana*. Molecular Plant. 2012;6(2):503–13.

[73] Du J, Huang Z, Wang B, Sun H, Chen C, Ling HQ, et al. SlbHLH068 interacts with FER to regulate the iron-deficiency response in tomato. Annals of Botany. 2015;116(1):23–34.

[74] Pires N, Dolan L. Origin and diversification of basic-helix-loop-helix proteins in plants. Molecular Biology and Evolution. 2010;27(4):862–74.

[75] Wang HY, Klatte M, Jakoby M, Baumlein H, Weisshaar B, Bauer P. Iron deficiency-mediated stress regulation of four subgroup Ib BHLH genes in *Arabidopsis thaliana*. Planta. 2007;226(4):897–908.

[76] Schuler M, Keller A, Backes C, Philippar K, Lenhof HP, Bauer P. Transcriptome analysis by GeneTrail revealed regulation of functional categories in response to alterations of iron homeostasis in *Arabidopsis thaliana*. BMC Plant Biology. 2011;11:87.

[77] Bauer P, Blondet E. Transcriptome analysis of ein3 eil1 mutants in response to iron deficiency. Plant Signaling & Behavior. 2011;6(11):1669–71.

[78] Maurer F, Naranjo Arcos MA, Bauer P. Responses of a triple mutant defective in three iron deficiency-induced Basic Helix-Loop-Helix genes of the subgroup Ib(2) to iron deficiency and salicylic acid. PLoS One. 2014;9(6):e99234.

[79] Ivanov R, Brumbarova T, Bauer P. Fitting into the harsh reality: regulation of iron-deficiency responses in dicotyledonous plants. Molecular Plant. 2012;5(1):27–42.

[80] Yang TJ, Lin WD, Schmidt W. Transcriptional profiling of the Arabidopsis iron deficiency response reveals conserved transition metal homeostasis networks. Plant Physiology. 2010;152(4):2130–41.

[81] Buckhout TJ, Yang TJ, Schmidt W. Early iron-deficiency-induced transcriptional changes in Arabidopsis roots as revealed by microarray analyses. BMC Genomics. 2009;10:147.

[82] Long TA, Tsukagoshi H, Busch W, Lahner B, Salt DE, Benfey PN. The bHLH transcription factor POPEYE regulates response to iron deficiency in Arabidopsis roots. The Plant Cell. 2010;22(7):2219–36.

[83] Kobayashi T, Nagasaka S, Senoura T, Itai RN, Nakanishi H, Nishizawa NK. Iron-binding haemerythrin RING ubiquitin ligases regulate plant iron responses and accumulation. Nature Communications. 2013;4:2792.

[84] Zhang J, Liu B, Li M, Feng D, Jin H, Wang P, et al. The bHLH transcription factor bHLH104 interacts with IAA-LEUCINE RESISTANT3 and modulates iron homeostasis in Arabidopsis. The Plant Cell. 2015;27(3):787–805.

[85] Palmer CM, Hindt MN, Schmidt H, Clemens S, Guerinot ML. MYB10 and MYB72 are required for growth under iron-limiting conditions. PLoS Genet. 2013;9(11):e1003953.

[86] Sivitz AB, Hermand V, Curie C, Vert G. Arabidopsis bHLH100 and bHLH101 control iron homeostasis via a FIT-independent pathway. PLoS ONE. 2012;7(9):e44843.

[87] Zamioudis C, Hanson J, Pieterse CM. beta-Glucosidase BGLU42 is a MYB72-dependent key regulator of rhizobacteria-induced systemic resistance and modulates iron deficiency responses in Arabidopsis roots. New Phytologist. 2014;204(2):368–79.

[88] Andrews N. Disorders of iron metabolism. The New England Journal of Medicine. 1999;341(26):1986–95.

[89] Conrad M, Umbreit J. Iron absorption and transport? An update. American Journal of Hematology. 2000;64(4):287–98.

[90] Lieu P, Heiskala M, Peterson P, Yang Y. The roles of iron in health and disease. Molecular Aspects of Medicine. 2001;22(1):1–87.

[91] Trumbo P, Yates AA, Schlicker S, Poos M. Dietary reference intakes. Journal of the American Dietetic Association. 2001;101(3):294–301.

[92] Baynes R, Bothwell T. Iron deficiency. Annual Review of Nutrition. 1990;10(1):133–48.

[93] Monsen ER, Hallberg L, Layrisse M, Hegsted DM, Cook JD, Mertz W, et al. Estimation of available dietary iron. The American Journal of Clinical Nutrition. 1978;31(1): 134–41.

[94] Hunt JR. Moving toward a plant-based diet: are iron and zinc at risk? Nutrition Reviews. 2002;60(5 Pt 1):127–34.

[95] Hunt JR. Bioavailability of iron, zinc, and other trace minerals from vegetarian diets. The American Journal of Clinical Nutrition. 2003;78(3 Suppl):633s–9s.

[96] Lopez HW, Leenhardt F, Coudray C, Remesy C. Minerals and phytic acid interactions: is it a real problem for human nutrition? International Journal of Food Science & Ttechnology. 2002;37(7):727–39.

[97] Petry N, Egli I, Zeder C, Walczyk T, Hurrell R. Polyphenols and phytic acid contribute to the low iron bioavailability from common beans in young women. The Journal of Nutrition. 2010;140(11):1977–82.

[98] Nielsen AV, Tetens I, Meyer AS. Potential of phytase-mediated iron release from cereal-based foods: a quantitative view. Nutrients. 2013;5(8):3074–98.

[99] Schlemmer U, Frolich W, Prieto RM, Grases F. Phytate in foods and significance for humans: food sources, intake, processing, bioavailability, protective role and analysis. Molecular Nutrition & Food Research. 2009;53 Suppl 2:S330–75.

[100] Gillooly M, Bothwell T, Torrance J, MacPhail A, Derman D, Bezwoda W, et al. The effects of organic acids, phytates and polyphenols on the absorption of iron from vegetables. British Journal of Nutrition. 1983;49(03):331–42.

[101] Hallberg L, Brune M, Rossander L. Iron absorption in man: ascorbic acid and dose-dependent inhibition by phytate. The American Journal of Clinical Nutrition. 1989;49(1):140–4.

[102] Graf E, Empson KL, Eaton JW. Phytic acid. A natural antioxidant. Journal of Biological Chemistry. 1987;262(24):11647–50.

[103] Hurrell R. Bioavailability of iron. European Journal of Clinical Nutrition. 1997;51(1):S4.

[104] Conrad ME, Umbreit JN. Iron absorption and transport-an update. American Journal of Hematology. 2000;64(4):287–98.

[105] Garcia-Casal MN, Layrisse M, Solano L, Baron MA, Arguello F, Llovera D, et al. Vitamin A and beta-carotene can improve nonheme iron absorption from rice, wheat and corn by humans. The Journal of Nutrition. 1998;128(3):646–50.

[106] Hurrell R, Egli I. Iron bioavailability and dietary reference values. The American Journal of Clinical Nutrition. 2010;91(5):1461S–7S.

[107] Lynch SR, Cook JD. Interaction of vitamin C and iron. Annals of the New York Academy of Sciences. 1980;355:32–44.

[108] Light W, Olson JS. Transmembrane movement of heme. Journal of Biological Chemistry. 1990;265(26):15623–31.

[109] Hunt JR. Dietary and physiological factors that affect the absorption and bioavailability of iron. International Journal for Vitamin and Nutrition Research. 2005;75(6):375–84.

[110] Shayeghi M, Latunde-Dada GO, Oakhill JS, Laftah AH, Takeuchi K, Halliday N, et al. Identification of an intestinal heme transporter. Cell. 2005;122(5):789–801.

[111] Qiu A, Jansen M, Sakaris A, Min SH, Chattopadhyay S, Tsai E, et al. Identification of an intestinal folate transporter and the molecular basis for hereditary folate malabsorption. Cell. 2006;127(5):917–28.

[112] Lieu PT, Heiskala M, Peterson PA, Yang Y. The roles of iron in health and disease. Molecular Aspects of Medicine. 2001;22(1–2):1–87.

[113] Wyman S, Simpson RJ, McKie AT, Sharp PA. Dcytb (Cybrd1) functions as both a ferric and a cupric reductase in vitro. FEBS Letters. 2008;582(13):1901–6.

[114] Mackenzie B, Ujwal ML, Chang MH, Romero MF, Hediger MA. Divalent metal-ion transporter DMT1 mediates both H+ -coupled Fe2+ transport and uncoupled fluxes. Pflugers Archiv. 2006;451(4):544–58.

[115] Zoller H, Theurl I, Koch R, Kaser A, Weiss G. Mechanisms of iron mediated regulation of the duodenal iron transporters divalent metal transporter 1 and ferroportin 1. Blood Cells, Molecules, and Diseases. 2002;29(3):488–97.

[116] Vulpe CD, Kuo YM, Murphy TL, Cowley L, Askwith C, Libina N, et al. Hephaestin, a ceruloplasmin homologue implicated in intestinal iron transport, is defective in the sla mouse. Nature Genetics. 1999;21(2):195–9.

[117] Lynch SR. Iron metabolism. Kraemer K, Zimmermann MB. Nutritional Anemia. 2007. p. 59–76. 3-906412-33-4.

[118] Irie S, Tavassoli M. Transferrin-mediated cellular iron uptake. The American Journal of the Medical Sciences. 1987;293(2):103–11.

[119] Goswami T, Andrews NC. Hereditary hemochromatosis protein, HFE, interaction with transferrin receptor 2 suggests a molecular mechanism for mammalian iron sensing. Journal of Biological Chemistry. 2006;281(39):28494–8.

[120] Bennett MJ, Lebrón JA, Bjorkman PJ. Crystal structure of the hereditary haemochromatosis protein HFE complexed with transferrin receptor. Nature. 2000;403(6765):46–53.

[121] Fleming RE, Ponka P. Iron overload in human disease. New England Journal of Medicine. 2012;366(4):348–59.

[122] Nicolas G, Chauvet C, Viatte L, Danan JL, Bigard X, Devaux I, et al. The gene encoding the iron regulatory peptide hepcidin is regulated by anemia, hypoxia, and inflammation. Journal of Clinical Investigation. 2002;110(7):1037–44.

[123] Pigeon C, Ilyin G, Courselaud B, Leroyer P, Turlin B, Brissot P, et al. A new mouse liver-specific gene, encoding a protein homologous to human antimicrobial peptide hepcidin, is overexpressed during iron overload. Journal of Biological Chemistry. 2001;276(11):7811–9.

[124] Mena NP, Esparza A, Tapia V, Valdes P, Nunez MT. Hepcidin inhibits apical iron uptake in intestinal cells. American Journal of Physiology-Gastrointestinal and Liver Physiology. 2008;294(1):G192–8.

[125] Brasse-Lagnel C, Karim Z, Letteron P, Bekri S, Bado A, Beaumont C. Intestinal DMT1 cotransporter is down-regulated by hepcidin via proteasome internalization and degradation. Gastroenterology. 2011;140(4):1261–71, e1.

[126] Nemeth E, Tuttle MS, Powelson J, Vaughn MB, Donovan A, Ward DM, et al. Hepcidin regulates cellular iron efflux by binding to ferroportin and inducing its internalization. Science (New York, NY). 2004;306(5704):2090–3.

[127] Delaby C, Pilard N, Goncalves AS, Beaumont C, Canonne-Hergaux F. Presence of the iron exporter ferroportin at the plasma membrane of macrophages is enhanced by iron loading and down-regulated by hepcidin. Blood. 2005;106(12):3979–84.

[128] Yamaji S, Sharp P, Ramesh B, Srai SK. Inhibition of iron transport across human intestinal epithelial cells by hepcidin. Blood. 2004;104(7):2178–80.

[129] Niederkofler V, Salie R, Arber S. Hemojuvelin is essential for dietary iron sensing, and its mutation leads to severe iron overload. The Journal of Clinical Investigation. 2005;115(8):2180–6.

[130] Wang CY, Meynard D, Lin HY. The role of TMPRSS6/matriptase-2 in iron regulation and anemia. Frontiers in Pharmacology. 2014;5:114.

[131] Gulec S, Anderson GJ, Collins JF. Mechanistic and regulatory aspects of intestinal iron absorption. American Journal of Physiology-Gastrointestinal and Liver Physiology. 2014;307(4):G397–409.

[132] Pietrangelo A. Hereditary hemochromatosis: pathogenesis, diagnosis, and treatment. Gastroenterology. 2010;139(2):393–408, e1–2.

[133] Halliwell B, Gutteridge JM. Oxygen toxicity, oxygen radicals, transition metals and disease. Biochemical Journal. 1984;219(1):1–14.

[134] Sies H. Oxidative stress. Michigan, 1985. p. 507. 9780126427608.

[135] Ahmad P, Sarwat M, Sharma S. Reactive Oxygen Species, Antioxidants and Signaling in Plants. Journal of Plant Biology. 2008;51(3):167–73.

[136] Turrens JF. Mitochondrial formation of reactive oxygen species. The Journal of Physiology. 2003;552(Pt 2):335–44.

[137] Apel K, Hirt H. Reactive oxygen species: metabolism, oxidative stress, and signal transduction. Annual Review of Plant Biology. 2004;55:373–99.

[138] Schreck R, Rieber P, Baeuerle PA. Reactive oxygen intermediates as apparently widely used messengers in the activation of the NF-kappa B transcription factor and HIV-1. The EMBO Journal. 1991;10(8):2247.

[139] Pahl HL, Baeuerle PA. Oxygen and the control of gene expression. Bioessays. 1994;16(7):497–502.

[140] Dalton TP, Shertzer HG, Puga A. Regulation of gene expression by reactive oxygen. Annual Review of Pharmacology and Toxicology. 1999;39(1):67–101.

[141] Dat J, Vandenabeele S, Vranová E, Van Montagu M, Inzé D, Van Breusegem F. Dual action of the active oxygen species during plant stress responses. Cellular and Molecular Life Sciences CMLS. 2000;57(5):779–95.

[142] Low P, Merida J. The oxidative burst in plant defense: function and signal transduction. Physiologia Plantarum. 1996;96(3):533–42.

[143] Fenton H. Oxidation of tartaric acid in presence of iron. Journal of the Chemical Society. 1894;65.

[144] Haber F, Weiss J, editors. The catalytic decomposition of hydrogen peroxide by iron salts. Proceedings of the Royal Society of London A: Mathematical, Physical and Engineering Sciences; 1934: The Royal Society.

[145] Kehrer JP. The Haber–Weiss reaction and mechanisms of toxicity. Toxicology. 2000;149(1):43–50.

[146] Halliwell B. Free radicals, antioxidants, and human disease: curiosity, cause, or consequence? The Lancet. 1994;344(8924):721–4.

[147] McLaren CE, Barton JC, Adams PC, Harris EL, Acton RT, Press N, et al. Hemochromatosis and Iron Overload Screening (HEIRS) study design for an evaluation of 100,000 primary care-based adults. The American Journal of the Medical Sciences. 2003;325(2): 53–62.

[148] Mittler R. Oxidative stress, antioxidants and stress tolerance. Trends in Plant Science. 2002;7(9):405–10.

[149] Schrader M, Fahimi HD. Mammalian peroxisomes and reactive oxygen species. Histochemistry and Cell Biology. 2004;122(4):383–93.

[150] Tripathy BC, Oelmuller R. Reactive oxygen species generation and signaling in plants. Plant Signaling & Behavior. 2012;7(12):1621–33.

[151] Halliwell B. Reactive species and antioxidants. Redox biology is a fundamental theme of aerobic life. Plant Physiology. 2006;141(2):312–22.

[152] Fridovich I. Superoxide radical and superoxide dismutases. Annual Review of Biochemistry. 1995;64(1):97–112.

[153] Alscher RG, Erturk N, Heath LS. Role of superoxide dismutases (SODs) in controlling oxidative stress in plants. Journal of Experimental Botany. 2002;53(372):1331–41.

[154] Noctor G, Mhamdi A, Chaouch S, Han Y, Neukermans J, Marquez-Garcia B, et al. Glutathione in plants: an integrated overview. Plant, Cell & Environment. 2012;35(2): 454–84.

[155] Eshdat Y, Holland D, Faltin Z, Ben-Hayyim G. Plant glutathione peroxidases. Physiologia Plantarum. 1997;100(2):234–40.

[156] Noctor G, Foyer CH. Ascorbate and glutathione: keeping active oxygen under control. Annual Review of Plant Biology. 1998;49(1):249–79.

[157] Signorelli S, Coitino EL, Borsani O, Monza J. Molecular mechanisms for the reaction between ()OH radicals and proline: insights on the role as reactive oxygen species scavenger in plant stress. The Journal of Physical Chemistry B. 2014;118(1):37–47.

[158] Kavi Kishor PB, Sreenivasulu N. Is proline accumulation per se correlated with stress tolerance or is proline homeostasis a more critical issue? Plant, Cell & Environment. 2014;37(2):300–11.

[159] Menden B, Kohlhoff M, Moerschbacher BM. Wheat cells accumulate a syringyl-rich lignin during the hypersensitive resistance response. Phytochemistry. 2007;68(4):513–20.

[160] Lange BM, Lapierre C, Sandermann Jr H. Elicitor-induced spruce stress lignin (structural similarity to early developmental lignins). Plant Physiology. 1995;108(3): 1277–87.

[161] Smit F, Dubery IA. Cell wall reinforcement in cotton hypocotyls in response to a *Verticillium dahliae* elicitor. Phytochemistry. 1997;44(5):811–5.

[162] Sattler SE, Funnell-Harris DL. Modifying lignin to improve bioenergy feedstocks: strengthening the barrier against pathogens? Frontiers in Plant Science. 2013;4.

[163] Elias PM. The skin barrier as an innate immune element. Seminars in Immunopathology. 2007;29(1):3–14.

[164] Akpek E, Gottsch J. Immune defense at the ocular surface. Eye. 2003;17(8):949–56.

[165] Corfield AP. Mucins: a biologically relevant glycan barrier in mucosal protection. Biochimica et Biophysica Acta. 2015;1850(1):236–52.

[166] Monaghan J, Zipfel C. Plant pattern recognition receptor complexes at the plasma membrane. Current Opinion in Plant Biology. 2012;15(4):349–57.

[167] Boller T, Felix G. A renaissance of elicitors: perception of microbe-associated molecular patterns and danger signals by pattern-recognition receptors. Annual Review of Plant Biology. 2009;60:379–406.

[168] Ranf S, Gisch N, Schaffer M, Illig T, Westphal L, Knirel YA, et al. A lectin S-domain receptor kinase mediates lipopolysaccharide sensing in *Arabidopsis thaliana*. Nature Immunology. 2015;16(4):426–33.

[169] Ramos HC, Rumbo M, Sirard JC. Bacterial flagellins: mediators of pathogenicity and host immune responses in mucosa. Trends in Microbiology. 2004;12(11):509–17.

[170] Willmann R, Lajunen HM, Erbs G, Newman M-A, Kolb D, Tsuda K, et al. Arabidopsis lysin-motif proteins LYM1 LYM3 CERK1 mediate bacterial peptidoglycan sensing and immunity to bacterial infection. Proceedings of the National Academy of Sciences. 2011;108(49):19824–9.

[171] Lu D, Wu S, Gao X, Zhang Y, Shan L, He P. A receptor-like cytoplasmic kinase, BIK1, associates with a flagellin receptor complex to initiate plant innate immunity. Proceedings of the National Academy of Sciences USA. 2010;107(1):496–501.

[172] Postel S, Kufner I, Beuter C, Mazzotta S, Schwedt A, Borlotti A, et al. The multifunctional leucine-rich repeat receptor kinase BAK1 is implicated in Arabidopsis development and immunity. European Journal of Cell Biology. 2010;89(2–3):169–74.

[173] Schwessinger B, Roux M, Kadota Y, Ntoukakis V, Sklenar J, Jones A, et al. Phosphorylation-dependent differential regulation of plant growth, cell death, and innate immunity by the regulatory receptor-like kinase BAK1. PLoS Genet. 2011;7(4):e1002046.

[174] Salmond GP, Reeves PJ. Membrance traffic wardens and protein secretion in Gram-negative bacteria. Trends in Biochemical Sciences. 1993;18(1):7–12.

[175] Thakur M, Sohal BS. Role of elicitors in inducing resistance in plants against pathogen infection: a review. ISRN Biochemistry. 2013;2013:762412.

[176] Abramovitch RB, Anderson JC, Martin GB. Bacterial elicitation and evasion of plant innate immunity. Nature Reviews Molecular Cell Biology. 2006;7(8):601–11.

[177] Alcazar R, Parker JE. The impact of temperature on balancing immune responsiveness and growth in Arabidopsis. Trends in Plant Science. 2011;16(12):666–75.

[178] Heath MC. Hypersensitive response-related death. Programmed Cell Death in Higher Plants: Springer; 2000. p. 77–90.

[179] Mishina TE, Zeier J. Pathogen-associated molecular pattern recognition rather than development of tissue necrosis contributes to bacterial induction of systemic acquired resistance in Arabidopsis. The Plant Journal. 2007;50(3):500–13.

[180] Shah J, Zeier J. Long-distance communication and signal amplification in systemic acquired resistance. Frontiers in Plant Science. 2013;4.

[181] Van Loon L, Van Strien E. The families of pathogenesis-related proteins, their activities, and comparative analysis of PR-1 type proteins. Physiological and Molecular Plant Pathology. 1999;55(2):85–97.

[182] Spoel SH, Dong X. How do plants achieve immunity? Defence without specialized immune cells. Nature Reviews Immunology. 2012;12(2):89–100.

[183] Pieterse CM, Zamioudis C, Berendsen RL, Weller DM, Van Wees SC, Bakker PA. Induced systemic resistance by beneficial microbes. Annual Review of Phytopathology. 2014;52:347–75.

[184] Zhang Y, Tessaro MJ, Lassner M, Li X. Knockout analysis of Arabidopsis transcription factors TGA2, TGA5, and TGA6 reveals their redundant and essential roles in systemic acquired resistance. The Plant Cell. 2003;15(11):2647–53.

[185] Spoel SH, Mou Z, Tada Y, Spivey NW, Genschik P, Dong X. Proteasome-mediated turnover of the transcription coactivator NPR1 plays dual roles in regulating plant immunity. Cell. 2009;137(5):860–72.

[186] Jaskiewicz M, Conrath U, Peterhansel C. Chromatin modification acts as a memory for systemic acquired resistance in the plant stress response. EMBO Reports. 2011;12(1):50–5.

[187] Conrath U, Beckers GJ, Flors V, García-Agustín P, Jakab G, Mauch F, et al. Priming: getting ready for battle. Molecular Plant-Microbe Interactions. 2006;19(10):1062–71.

[188] Beckers GJ, Jaskiewicz M, Liu Y, Underwood WR, He SY, Zhang S, et al. Mitogen-activated protein kinases 3 and 6 are required for full priming of stress responses in *Arabidopsis thaliana*. The Plant Cell. 2009;21(3):944–53.

[189] Luna E, Bruce TJ, Roberts MR, Flors V, Ton J. Next-generation systemic acquired resistance. Plant Physiology. 2012;158(2):844–53.

[190] Slaughter A, Daniel X, Flors V, Luna E, Hohn B, Mauch-Mani B. Descendants of primed Arabidopsis plants exhibit resistance to biotic stress. Plant Physiology. 2012;158(2):835–43.

[191] Lucht JM, Mauch-Mani B, Steiner HY, Metraux JP, Ryals J, Hohn B. Pathogen stress increases somatic recombination frequency in Arabidopsis. Nature Genetics. 2002;30(3):311–4.

[192] Barton GM. A calculated response: control of inflammation by the innate immune system. The Journal of Clinical Investigation. 2008;118(2):413–20.

[193] Banchereau J, Steinman RM. Dendritic cells and the control of immunity. Nature. 1998;392(6673):245–52.

[194] Holgate S. The role of mast cells and basophils in inflammation. Clinical and Experimental Allergy: Journal of the British Society for Allergy and Clinical Immunology. 2000;30:28–32.

[195] Guermonprez P, Valladeau J, Zitvogel L, Thery C, Amigorena S. Antigen presentation and T cell stimulation by dendritic cells. Annual Review of Immunology. 2002;20:621–67.

[196] Eichhorn H, Lessing F, Winterberg B, Schirawski J, Kamper J, Muller P, et al. A ferroxidation/permeation iron uptake system is required for virulence in *Ustilago maydis*. The Plant Cell. 2006;18(11):3332–45.

[197] Oide S, Moeder W, Krasnoff S, Gibson D, Haas H, Yoshioka K, et al. NPS6, encoding a nonribosomal peptide synthetase involved in siderophore-mediated iron metabolism, is a conserved virulence determinant of plant pathogenic ascomycetes. The Plant Cell. 2006;18(10):2836–53.

[198] Hwang LH, Seth E, Gilmore SA, Sil A. SRE1 regulates iron-dependent and -independent pathways in the fungal pathogen Histoplasma capsulatum. Eukaryotic Cell. 2012;11(1):16–25.

[199] Dellagi A, Rigault M, Segond D, Roux C, Kraepiel Y, Cellier F, et al. Siderophore-mediated upregulation of Arabidopsis ferritin expression in response to Erwinia chrysanthemi infection. The Plant Journal. 2005;43(2):262–72.

[200] Liu G, Greenshields DL, Sammynaiken R, Hirji RN, Selvaraj G, Wei Y. Targeted alterations in iron homeostasis underlie plant defense responses. Journal of Cell Science. 2007;120(4):596–605.

[201] Ganz T, Nemeth E. Iron homeostasis in host defence and inflammation. Nature Reviews Immunology. 2015;15(8):500–10.

[202] Schaible UE, Kaufmann SH. Iron and microbial infection. Nature Reviews Microbiology. 2004;2(12):946–53.

[203] Ratledge C, Dover LG. Iron metabolism in pathogenic bacteria. Annual Reviews in Microbiology. 2000;54(1):881–941.

[204] Chet I, Ordentlich A, Shapira R, Oppenheim A. Mechanisms of biocontrol of soil-borne plant pathogens by rhizobacteria. Plant and Soil. 1990;129(1):85–92.

[205] Masalha J, Kosegarten H, Elmaci Ö, Mengel K. The central role of microbial activity for iron acquisition in maize and sunflower. Biology and Fertility of Soils. 2000;30(5–6): 433–9.

[206] Pii Y, Borruso L, Brusetti L, Crecchio C, Cesco S, Mimmo T. The interaction between iron nutrition, plant species and soil type shapes the rhizosphere microbiome. Plant Physiology and Biochemistry. 2016;99:39–48.

[207] Iavicoli A, Boutet E, Buchala A, Metraux JP. Induced systemic resistance in *Arabidopsis thaliana* in response to root inoculation with Pseudomonas fluorescens CHA0. Molecular plant-microbe interactions: MPMI. 2003;16(10):851–8.

[208] Bakker PA, Pieterse CM, van Loon LC. Induced systemic resistance by fluorescent seudomonas spp. Phytopathology. 2007;97(2):239–43.

[209] Zamioudis C, Korteland J, Van Pelt JA, van Hamersveld M, Dombrowski N, Bai Y, et al. Rhizobacterial volatiles and photosynthesis-related signals coordinate MYB72 expression in Arabidopsis roots during onset of induced systemic resistance and iron-deficiency responses. The Plant Journal. 2015;84(2):309–22.

[210] Aznar A, Chen NW, Rigault M, Riache N, Joseph D, Desmaele D, et al. Scavenging iron: a novel mechanism of plant immunity activation by microbial siderophores. Plant Physiology. 2014;164(4):2167–83.

[211] Cohen YR. β-aminobutyric acid-induced resistance against plant pathogens. Plant Disease. 2002;86(5):448–57.

[212] Ton J, Jakab G, Toquin V, Flors V, Iavicoli A, Maeder MN, et al. Dissecting the beta-aminobutyric acid-induced priming phenomenon in Arabidopsis. The Plant Cell. 2005;17(3):987–99.

[213] Jakab G, Cottier V, Toquin V, Rigoli G, Zimmerli L, Métraux J-P, et al. β-Aminobutyric acid-induced resistance in plants. European Journal of Plant Pathology. 2001;107(1): 29–37.

[214] Koen E, Trapet P, Brule D, Kulik A, Klinguer A, Atauri-Miranda L, et al. Beta-Aminobutyric acid (BABA)-induced resistance in *Arabidopsis thaliana*: link with iron homeostasis. Molecular Plant-microbe Interactions: MPMI. 2014;27(11):1226–40.

[215] Kang HG, Singh KB. Characterization of salicylic acid-responsive, arabidopsis Dof domain proteins: overexpression of OBP3 leads to growth defects. The Plant Journal. 2000;21(4):329–39.

[216] Ward JM, Cufr CA, Denzel MA, Neff MM. The Dof transcription factor OBP3 modulates phytochrome and cryptochrome signaling in Arabidopsis. The Plant Cell. 2005;17(2):475–85.

[217] Kang HG, Foley RC, Onate-Sanchez L, Lin C, Singh KB. Target genes for OBP3, a Dof transcription factor, include novel basic helix-loop-helix domain proteins inducible by salicylic acid. The Plant Journal. 2003;35(3):362–72.

[218] Gregorio GB, Senadhira D, Htut H, Graham RD. Breeding for trace mineral density in rice. Food and Nutrition Bulletin. 2000;21(4):382–6.

[219] Haas JD, Beard JL, Murray-Kolb LE, del Mundo AM, Felix A, Gregorio GB. Iron-biofortified rice improves the iron stores of nonanemic Filipino women. The Journal of Nutrition. 2005;135(12):2823–30.

[220] Suzuki M, Morikawa KC, Nakanishi H, Takahashi M, Saigusa M, Mori S, et al. Transgenic rice lines that include barley genes have increased tolerance to low iron availability in a calcareous paddy soil. Soil Science and Plant Nutrition. 2008;54(1):77–85.

[221] Ishimaru Y, Kim S, Tsukamoto T, Oki H, Kobayashi T, Watanabe S, et al. Mutational reconstructed ferric chelate reductase confers enhanced tolerance in rice to iron deficiency in calcareous soil. Proceedings of the National Academy of Sciences USA. 2007;104(18):7373–8.

[222] Vert G, Grotz N, Dedaldechamp F, Gaymard F, Guerinot ML, Briat JF, et al. IRT1, an Arabidopsis transporter essential for iron uptake from the soil and for plant growth. The Plant Cell. 2002;14(6):1223–33.

[223] Sperotto RA, Ricachenevsky FK, Waldow Vde A, Fett JP. Iron biofortification in rice: it's a long way to the top. Plant Science. 2012;190:24–39.

[224] Drakakaki G, Marcel S, Glahn RP, Lund EK, Pariagh S, Fischer R, et al. Endosperm-specific co-expression of recombinant soybean ferritin and Aspergillus phytase in maize results in significant increases in the levels of bioavailable iron. Plant Molecular Biology. 2005;59(6):869–80.

[225] Lucca P, Hurrell R, Potrykus I. Genetic engineering approaches to improve the bioavailability and the level of iron in rice grains. Theoretical and Applied Genetics. 2001;102(2–3):392–7.

[226] Goto F, Yoshihara T, Saiki H. Iron accumulation and enhanced growth in transgenic lettuce plants expressing the iron-binding protein ferritin. Theoretical and Applied Genetics. 2000;100(5):658–64.

[227] Punja ZK. Genetic engineering of plants to enhance resistance to fungal pathogens—a review of progress and future prospects. Canadian Journal of Plant Pathology. 2001;23(3):216–35.

[228] Singh NK, Bracker CA, Hasegawa PM, Handa AK, Buckel S, Hermodson MA, et al. Characterization of osmotin A thaumatin-like protein associated with osmotic adaptation in plant cells. Plant Physiology. 1987;85(2):529–36.

[229] Woloshuk CP, Meulenhoff JS, Sela-Buurlage M, Van den Elzen P, Cornelissen B. Pathogen-induced proteins with inhibitory activity toward Phytophthora infestans. The Plant Cell. 1991;3(6):619–28.

[230] Liu D, Raghothama KG, Hasegawa PM, Bressan RA. Osmotin overexpression in potato delays development of disease symptoms. Proceedings of the National Academy of Sciences. 1994;91(5):1888–92.

[231] Ganz T, Lehrer RI. Defensins. Current Opinion in Immunology. 1994;6(4):584–9.

[232] Broekaert WF, Terras F, Cammue B, Osborn RW. Plant defensins: novel antimicrobial peptides as components of the host defense system. Plant Physiology. 1995;108(4):1353.

[233] Gao A-G, Hakimi SM, Mittanck CA, Wu Y, Woerner BM, Stark DM, et al. Fungal pathogen protection in potato by expression of a plant defensin peptide. Nature Biotechnology. 2000;18(12):1307–10.

[234] Georgopapadakou NH, Tkacz JS. The fungal cell wall as a drug target. Trends in Microbiology. 1995;3(3):98–104.

[235] Bolar JP, Norelli JL, Wong K-W, Hayes CK, Harman GE, Aldwinckle HS. Expression of endochitinase from Trichoderma harzianum in transgenic apple increases resistance to apple scab and reduces vigor. Phytopathology. 2000;90(1):72–7.

[236] Wróbel-Kwiatkowska M, Lorenc-Kukula K, Starzycki M, Oszmiański J, Kepczyńska E, Szopa J. Expression of β-1, 3-glucanase in flax causes increased resistance to fungi. Physiological and Molecular Plant Pathology. 2004;65(5):245–56.

[237] Ezaki B, Gardner RC, Ezaki Y, Matsumoto H. Expression of aluminum-induced genes in transgenic Arabidopsis plants can ameliorate aluminum stress and/or oxidative stress. Plant Physiology. 2000;122(3):657–66.

[238] Kwon S, Jeong Y, Lee H, Kim J, Cho K, Allen R, et al. Enhanced tolerances of transgenic tobacco plants expressing both superoxide dismutase and ascorbate peroxidase in chloroplasts against methyl viologen-mediated oxidative stress. Plant, Cell & Environment. 2002;25(7):873–82.

Chapter 5

An Integrated Approach to Iron Deficiency Anemia

Halima Nazar and Khan Usmanghani

Additional information is available at the end of the chapter

http://dx.doi.org/10.5772/63932

Abstract

Iron deficiency is a common nutritional disorder in developing countries and contributes significantly to reduced work productivity and economic output as well as to increased morbidity and mortality. There are well-established biochemical tests for assessing iron status in developed countries. However, cost and interference from infectious conditions make it difficult to assess iron status in many developing country settings. Examination of the hemoglobin distribution in the population and assessment of the hemoglobin response to supplementation are alternative approaches to define iron status and the nature of anemia. Prevention and control of iron deficiency requires the combined approach of dietary improvement, fortification of a common staple food when feasible, and appropriate iron supplementation for infants and pregnant women. In all these intervention activities, operational research is needed to improve effectiveness. In addition, controlling iron deficiency requires coordination with other nutrition and primary health care programs as part of an integrated approach to improved health and nutrition of the population. A randomized, controlled double-blind clinical trial was conducted to compare the efficacy and safety of herbal medicinal treatment syrup Sharbat-a-Folad versus syrup Ferplex for the treatment of iron deficiency anemia (IDA).

Keywords: iron deficiency, herbal medicine, anemia

1. Introduction

Iron deficiency anemia (IDA) is the most common nutritional deficiency worldwide. It can cause reduced work capacity in adults [1] and impact motor and mental development in children and adolescents [2]. There is some evidence that iron deficiency without anemia affects cognition in adolescent girls [3] and causes fatigue in adult women [4]. IDA may affect visual and auditory functioning and is weakly associated with poor cognitive development in children.

The term "anemia" is used for a group of conditions in which the number of red blood cells in the blood is lower than normal, or the red blood cells do not have enough hemoglobin. The estimates of the prevalence of anemia vary widely and accurate data are often lacking, and it can be assumed that significant proportions of young children and women of childbearing age are anemic [5, 6].

Iron deficiency results when iron demand by the body is not met by iron absorption from the diet. Thus, patients with IDA presenting in primary care may have inadequate dietary intake, hampered absorption, or physiologic losses in a woman of reproductive age. It also could be a sign of blood loss, known or occult. IDA is never an end diagnosis; the work-up is not complete until the reason for IDA is known.

The risk factors associated with IDA includes the following: low socioeconomic status, race as black women have a lower mean hemoglobin and a wider standard deviation than white women, inadequate dietary intake or parity, suggesting that there may be an unidentified, possibly racial factor predisposing these women to iron deficiency [7].

Anemia cannot be reliably diagnosed by clinical presentation. Fatigue, the most common reason to check hemoglobin, was caused by anemia in only one out of 52 patients in a primary care practice [8]. In a hospital setting, pallor predicted anemia with a likelihood ratio (LR) of 4.5. However, the absence of pallor was less helpful in ruling out anemia, giving an LR of 0.6 even when anemia was defined as less than 9 g per dL (90 g per L), a lower diagnostic level than that of the World Health Organization (WHO) or the Centers for Disease Control and Prevention (CDC) [9]. Other classic symptoms such as koilonychia (spoon nails), glossitis, or dysphagia are not common in the developed world [10].

The diagnosis of IDA requires that a patient be anemic and show laboratory evidence of iron deficiency. Red blood cells in IDA are usually described as being microcytic (i.e., mean corpuscular volume less than 80 μm^3 [80 fL]) and hypochromic; however, the manifestation of iron deficiency occurs in several stages [11]. Patients with a serum ferritin concentration less than 25 ng per mL (25 mcg per L) have a very high probability of being iron deficient. The most accurate initial diagnostic test for IDA is the serum ferritin measurement. Serum ferritin values greater than 100 ng per mL (100 mcg per L) indicate adequate iron stores and a low likelihood ratio of IDA [12]. In some populations, such as those with inflammatory disease or cirrhosis, these tests must be interpreted slightly differently because ferritin is an acute-phase reactant. Cutoffs for abnormality in these patients generally are higher [13].

Another laboratory change that occurs in patients with IDA is an increase in the iron-carrying protein transferrin. The amount of iron available to bind to this molecule is reduced, causing a decrease in the transferrin saturation and an increase in the total iron-binding capacity. The serum transferrin receptor assay is a newer approach to measuring iron status at the cellular level. Increased levels are found in patients with IDA, and normal levels are found in patients with anemia of chronic disease [14].

The treatment arms were chosen by block randomization in batches of eight using computer-generated random numbers to assign women to one of the four combinations of trial intervention. Block randomization was chosen to ensure that the experimental groups would not become unbalanced if the rate of recruitment at sites differed greatly. A research fellow with no other role in the project is overseeing the labeling and packing of all the trial medications and holding the randomization schedule until the code is broken.

1.1. Level of significance

This is the set standard to decide the cutoff value between treatment groups when comparing the two groups. If the results are significant at this set level ($\alpha = 0.05$), the null hypothesis will be rejected.

2. Patients, materials, and methods

2.1. Study design

The study was based on an experimental, randomized double-blind clinical trial. The study had been conducted according to principles of good clinical practice (i.e., an informed consent was obtained before enrollment and proper history and clinical examination were recorded on each follow-up), and the study was carried out during May 2003 to June 2004. A randomized double-blind experimental design was employed to test the hypotheses; therefore, by manipulating the independent variables (efficacy, side effects), any effects on dependent variable (herbal and allopathic treatment) could be monitored.

2.2. Patients

The study was conducted on 50 patients aged 12–40 years who were attending gynecological outpatient visits in Shifa-ul-Mulk Memorial Hospital.

2.3. Setting

The study was conducted in the Department of Gynecology and Obstetrics at Shifa-ul-Mulk Memorial Hospital for Eastern Medicine at Hamdard University in Karachi.

2.4. Sample selection

In this study, only the patients selectively enrolled were diagnosed with IDA through clinical history and laboratory investigations were enrolled. Diagnosis of IDA was based on the typical signs and symptoms and laboratory finding. Complete blood picture (CBC), hemoglobin, erythrocyte sedimentation rate (ESR), and urine (routine and microscopic) tests were also performed.

Recruiting GPs identify eligible women in their clinical practice and invite them to consider participation in the trial, after provision of sufficient information to make an informed decision. Women who meet the eligibility criteria and agree to participate are required to give written informed consent. Recruiting GPs also obtain demographic and relevant past and current medical data, particularly data relevant to risk factors for developing iron deficiency anemia. The interventions being tested are as follows:

(i) Control group received allopathic treatment (syrup Ferplex two teaspoons for seven days), (ii) and the test group received the herbal medicine (two teaspoons of syrup Foulad for seven days). All participants were observed for three follow-up visits over the course of treatment until they improved.

Blood samples were collected for CBC when the clinical picture shows the complete improvement to access the efficacy of the trial and confirmation and improvement in hemoglobin status. Same parameters were followed in control group.

2.5. Assessment

The normal hemoglobin count in reference to age set standard of WHO, blood morphology, was used as the primary outcome of the study. Secondary outcomes included the total symptoms score, the global assessment of the treatment by the investigator, and the women safety of the drugs and severity of adverse events at each follow-up visit. The relationship of each event to the study drug was also assessed. The safety outcome measure was the incidence of treatment-emergent adverse events in both groups. A blood specimen for routine CBC was obtained prior to the treatment.

2.6. Inclusion criteria

Persons may be included in the trial, if they meet the following criteria:

- Female patients between the aged of 12–40 years suffering from IDA.
- All socioeconomical classes were included.
- Verbal consent and willingness to participate in all scheduled study visits and tests.
- Patients suffering from IDA.
- Patients living in Karachi, Pakistan.

2.7. Exclusion criteria

Patients were excluded, if they have any of the following criteria:

- Patients having other associated pathologies such as uncontrolled diabetes, hypertension, liver disorders, etc.
- Patients having other types of anemia such as protein deficiency anemia, pernicious anemia, sickle cell anemia, and thalassaemia.
- Known cases of iron therapy failure.
- Patients suffering from iron deficiency in secondary to malignancies.
- Patient belonging to any area outside Karachi because of intrinsic difficulty to follow up.

http://dx.doi.org/10.5772/63932

3. Results

The present study is to investigate formulated herbal medicine syrup Foulad for the treatment of IDA. The clinical screening of hematopoietic activity between Foulad and Ferplex was carried out to determine the efficacy and side effects towards off this malaise. These evaluations were based on clinical and laboratory findings so as to ascertain the rate of improvement in hemoglobin. In this study, a total of 50 patients were initially randomized and screened, the intent-to-treat population enrolled. The patients were evenly distributed to test or control group with the ratio of 1:1, that is, 25 in each group. This loss was distributed evenly between the treatment groups. The demographic and baseline characteristics of the intent-to-treat group were comparable for the herbal medicine and allopathic medicine treatment.

3.1. Patient characteristics

There were no significant differences in the mean age (26.12 ± 7.92 vs. 26.48 ± 4.75 test and control group, respectively, (**Table 1**) values between the treatment groups at the start of the clinical trial. All the patients were distributed in five-class interval ranging from age 12 to 40 years. The mean age of the married women's was 32.0 versus 32.46 and mean age of the single patients was 19.75 versus 20.0.

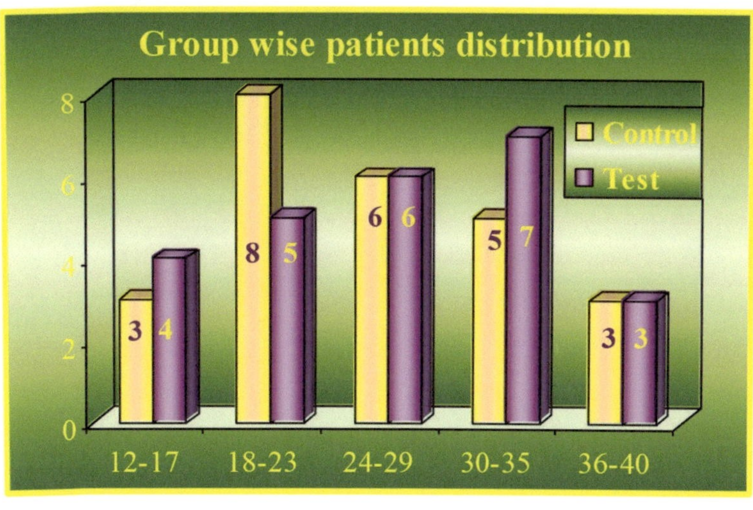

Marital status	Treatment group	Mean	Number (n)	Standard deviation	Sum
Married	Control	32.46	13	4.75	422
	Test	32.00	13	4.98	416
	Total	32.23	26	4.78	838

Marital status	Treatment group	Mean	Number (n)	Standard deviation	Sum
Single	Control	20.00	12	3.86	240
	Test	19.75	12	5.17	237
	Total	19.88	24	4.47	477
Total	Control	26.48	25	7.65	662
	Test	26.12	25	7.98	653
	Total	26.30	50	7.74	1315

Table 1. Marital status by treatment group.

3.2. Treatment assignment and follow-up

All subjects were clinically studied and completed assigned therapy during the period May 2001 to June 2004. Results presented below represent an intention-to-treat analysis, as stipulated by this study protocol. Baseline patient characteristics for all study variables were balanced among treatment groups (**Table 2**).

Anemia history and examination at baseline		Treatment group		Total n = 50	*p*value
		Control	Test		
Severity of anemia	Mild	10	7	17	0.650
	Moderate	12	15	27	
	Severe	3	3	6	
Symptoms of anemia	Asymptomatic	5	2	7	0.613
	Fatigue	2	5	7	
	Reduced concentration	0	1	1	
	Loss of appetite	3	1	4	
	Pica	1	2	3	
Signs of anemia	Koilonychia	1	1	2	
	Brittle nails	1	0	1	
	Pallor	9	11	20	
	Tachycardia	3	2	5	
Causes of anemia	Blood loss	14	13	27	0.990
	Decreased iron utilization	6	7	13	
	Dietary inadequacy	4	4	8	
	Malabsorption	1	1	2	

Table 2. Baseline demographic variables.

3.3. Baseline demographic variables

The baseline pretreatment analyses of IDA history and examination were performed. The clinical evaluation proforma of IDA was filled at the time of enrollment in both treatment groups. Patient's baseline demographic variables for IDA history and general physical examination were summarized for each treatment group. As depicted in **Table 2**, patient characteristics were equally balanced between the test and control groups. The two treatment groups did not differ significantly (all $p < 0.05$) from each other at any time point. The most common IDA symptom was pallor 44.0% in allopathic treated and 36.0% in herbal-treated patients. Whereas the common cause of IDA noted in this trial was the blood loss 56% in allopathic treated and 52% in herbal-treated patients.

3.4. Baseline severity of anemia

The assessment of severity of IDA at the time of enrollment exhibited following results test group versus control group, mild anemia 7 patients (28%) versus 10 (40%), moderate anemia 15 patients (60%) versus 12 (48%), severe anemia 3 patients (12%) versus 3 (12%) patients noticed in the both treatment groups. Baseline severity of anemia did not differ between the two groups. Comparative analysis of the baseline data using chi-square test confirms that there were no baseline differences among the treatment group as evident from p values in **Table 2.**

3.5. Effects of therapy on hemoglobin status

Hemoglobin increased dramatically in both treatment groups after therapy (as dissipated in graph 2). The rates of the improvement in hemoglobin concentration were higher in the herbal treatment group at all times after treatment and it revealed that the efficacy of herbal treatment is as superior to allopathic treatment ($p = 0.001$). The total duration of treatment was 4 weeks in both treatment groups. The clinical success rates on the basis of self-assessment of patient regression of complaints and physician examination on follow-up were more effective in test group (as in graph 2).

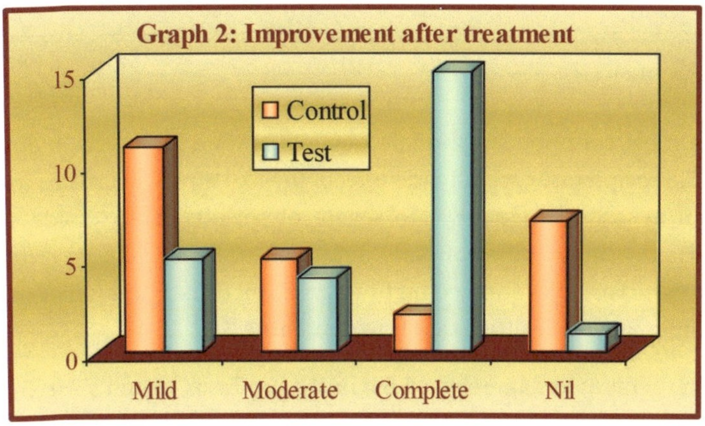

Clinical failures or no significant improvement in hemoglobin after treatment occurred in 1/25 patients (4%) receiving herbal medicine and in 7/25 patients (28%) receiving allopathic medicines (graph 2). Clinical success rates for those with mild-to-moderate infections and those with severe infections were higher in test treatment group. For the overall, clinical success was observed in 15/25 patients (60%) of cases in herbal-treated patients and in 2/25 (8%) of cases in allopathic-treated patients.

The overall evaluation was mainly based on the efficacy of drugs in reducing anemia in terms of both objective and subjective symptoms. The syrup Foulad produced a better result than allopathic medicines, which showed an overall cure rate of 60% versus 8% significantly effective as confirmed by chi-square test and the test treatment has superior efficacy than control treatment ($p = 0.001$).

3.6. Safety evaluations

All the patients enrolled in the study were evaluated for safety. Adverse effects observed after administration of medicine are summarized in **Table 3**. The majority of adverse events were assessed as mild in severity. Adverse events categorized by the physician (researcher) as possibly or definitely drug related were reported in 3/25 patients (12%) receiving herbal medicine and in 15/25 patients (60%) given allopathic medicine.

Observed Side effects	**Treatment group**		**Total**	***p* value**
	Control	**Test**		
Constipation	2	1	3	0.010
Diarrhea	3	1	4	
Nausea	9	1	10	
Vomiting	1	0	1	
No complaints	10	22	32	
Total patients	25	25	5	

Table 3. Side effects on patient's self-assessment.

Nausea was the most common drug-related events among allopathic medicine (36%) and herbal group (4%) recipients. Overall side effects ($p = 0.010$) were greater in control-treated participants than in test participants. No severe or serious adverse side effects were observed that interfere with activities of daily living. Comparison of data recorded by participants relating to these variables showed highly significant differences between test and control groups for measurements side effects as shown in **Table 3**.

Consequently, the generated data rejected the null hypothesis (when $p < 0.05$); hence, the null hypothesis was rejected on the basis of statistical findings in regard to efficacy and safety.

4. Discussion

Iron deficiency anemia (IDA) is most often a polysymptomatic disease. The use of allopathic drug combination has been considered as one of the effective therapy. But this is not feasible, as besides being cost prohibitive, they are not without side effects. The herbal formulation syrup Foulad contains herbs, which are known for its wide range of clinical use in indigenous medicine. It has been proved that these herbs exert profound activity for the improvement of hemoglobin percentage.

This unicenter trial demonstrated that herbal medicine was more effective in the management of patients with IDA. Herbal treatment resulted in a 60% clinical cure or improvement rate, which is superior to that achieved with allopathic therapy 8% clinical success rate. The response rate of hemoglobin improvement status before and after treatment suggests that syrup Foulad has the higher efficacy as allopathic medicine ($p > 0.001$).

In light of study that the authors have presented, it is concluded that phytomedicine administered under a randomized double-blind trial is exhibiting desirable effects with a profound margin of safety. The plus point is that the formulations are absolutely cost-effective and have shown promising results even in surveillance studies. The spectrum of herbal medicine has been widening following the modalities of integrated medicine like our eastern system of medicine.

The medicinal word is switching over to alternative medicine, including herbal medicine especially South Asia, due to its tremendous potential that is being confirmed by current researches. In this unicenter study, syrup Foulad was well tolerated and had a rate of drug-related adverse events less than to that of patients treated with syrup Ferplex ($p = 0.010$). Mild nausea was the most commonly reported adverse events in test treatment groups and control group.

5. Conclusion

Based on the statistical result of present clinical trial, it can be concluded that

- a comparative evaluation of the IDA treatment by Sharbat Foulad vis-a-vis the syrup Ferplex differs in treatment response, the herbal medicine is superior to allopathic medication.
- there was less untoward manifestation associated with the use of syrup Foulad and this is found a good acceptability by most of the treated patients. Syrup Foulad has added the benefit of safety.

Author details

Halima Nazar[1*] and Khan Usmanghani[2]

*Address all correspondence to: halimanazar76@gmail.com

1 Business Development Department, Rafay Laboratories Private Limited, Karachi, Pakistan.

2 Research and Development Department, Herbion Pakistan Private Limited, Karachi, Pakistan

References

[1] Haas JD, Brownlie T, Iron deficiency and reduced work capacity: a critical review of the research to determine a causal relationship, *J Nutr* 2001;131(2 suppl):676S–88S; discussion 688S-90S.

[2] Halterman JS, Kaczorowski JM, Aligne CA, Auinger P, Szilagyi PG, Iron deficiency and cognitive achievement among school-aged children and adolescents in the United States, *Pediatrics* 2001;107:1381–6.

[3] Algarin C, Peirano P, Garrido M, Pizarro F, Lozoff B, Iron deficiency anemia in infancy: long-lasting effects on auditory and visual system functioning, *Pediatr Res* 2003;53:217–23.

[4] Verdon F, Burnand B, Stubi CL, Bonard C, Graff M, Michaud A, Iron supplementation for unexplained fatigue in non-anaemic women: double blind randomized placebo controlled trial, *BMJ* 2003;326:1124.

[5] Aftab Saeed, Director Hamdard Research institute of unani Medicine hrium@live.com. Iron deficiency anaemia: assessment, prevention and control. 2001, Geneva, World Health Organization, 2001 (document WHO/NHD/01.3).

[6] Focusing on anaemia: towards an integrated approach for effective anaemia control: joint statement by the World Health Organization and the United Nations, Children's Fund, 2004, Geneva, World Health Organization.

[7] Ramakrishnan U, Frith-Terhune A, Cogswell M, Kettel Khan L, Dietary intake does not account for differences in low iron stores among Mexican American and non-Hispanic white women: Third National Health and Nutrition Examination Survey, 1988–1994, *J Nutr* 2002;132:996–1001.

[8] Elnicki DM, Shockcor WT, Brick JE, Beynon D, Evaluating the complaint of fatigue in primary care: diagnoses and outcomes, *Am J Med* 1992;93:303–6.

[9] Sheth TN, Choudhry NK, Bowes M, Detsky AS, The relation of conjunctival pallor to the presence of anemia, *J Gen Intern Med* 1997;12:102–6.

[10] Cook JD, Diagnosis and management of iron-deficiency anaemia, *Best Pract Res Clin Haematol* 2005;18:319–32.

[11] Zanella A, Gridelli L, Berzuini A, Colottie MT, Mozzi F, Milani S, Sensitivity and predictive value of serum ferritin and free erythrocyte protoporphyrin for iron deficiency, *J Lab Clin Med* 1989;113:73–8.

[12] Guyatt GH, Oxman AD, Ali M, Willan A, McIlroy W, Patterson C, Laboratory diagnosis of iron-deficiency anemia: an overview [published correction appears in J Gen Intern Med 1992;7:423], *J Gen Intern Med* 1992;7:145–53.

[13] Intragumtornchai T, Rojnukkarin P, Swasdikul D, Israsena S, The role of serum ferritin in the diagnosis of iron deficiency anaemia in patients with liver cirrhosis, *J Intern Med* 1998;243:233–41.

[14] Cook JD, Flowers CH, Skikne BS. The quantitative assessment of body iron. Blood. 2003;101:3359–3364.

Chapter 6

Anemia During Pregnancy

Ishag Adam and Abdelaziem A. Ali

Additional information is available at the end of the chapter

http://dx.doi.org/10.5772/63211

Abstract

Anemia during pregnancy is a considerable health problem, with around two-fifths of pregnant women worldwide being anemic. Many gynecological and infectious diseases are predisposing factors for anemia during pregnancy. Anemia during pregnancy—especially the severe form—can lead to various maternal and perinatal adverse effects such as preterm labor, low birth weight, and intrauterine fetal death. It is one of the leading causes of maternal mortality. Therefore, preventive measures are needed if anemia and its adverse effects are to be prevented. Iron and folic acid supplements are the cornerstone for the prevention of anemia during pregnancy and one of the earliest preventive measures adopted in antenatal care. Other measures to prevent anemia during pregnancy include the fortification of principle foods with iron, increasing health and nutritional awareness, combating parasitic infections, and improvement in sanitation. There is a controversy concerning the benefit of other elements such as zinc, copper, and magnesium, so the use of these elements is not widely adopted for the prevention of anemia.

Keywords: anemia, pregnancy, prevention, treatment, adverse effects

1. Introduction

Anemia in pregnancy is a major public health problem, where it has been estimated that 41.8% of pregnant women worldwide are anemic [1]. The majority (at least half) of this burden is due to iron deficiency [2]. However, there is a significant variation in prevalence of anemia, both within and between countries. Because of physiological changes during pregnancy, pregnant women are at higher risk of anemia and in particular iron deficiency anemia, which is the most common type of anemia during pregnancy. Hematological changes during pregnancy, especially expansion of blood volume, often confuse the diagnosis of anemia and its treatment. Because of increased iron and folic acid demands during pregnancy, pregnant women are more susceptible to develop anemia. Moreover, pregnant women are more susceptible to the other types of anemia that affect other women of childbearing age such as hereditary anemia, sickle cell disease and aplastic anemia. Anemia during pregnancy, particularly the severe form, is associated with increased maternal morbidity and mortality and contributes to 20% of the maternal mortality in Africa [2–5]. Anemia in pregnancy is associated with negative consequences for both the woman and neonate. Therefore, great effort is needed to develop/reassess and implement programs to control and prevent anemia during pregnancy.

2. Physiological changes during pregnancy related to anemia

During pregnancy, there is a considerable increase in plasma volume, which increases by 40–45% above the level of nonpregnant women. The blood volume expands by 15% compared with non-pregnancy levels. The disproportionate amount of blood creates the physiological and dilution anemia during pregnancy [5, 6]. However, these changes are of great importance and may protect pregnant woman against supine hypotension, guard against the adverse effects of the expectant blood loss during labor, and meet the demand for increased blood flow to the uterus and fetus [5, 7–9]. Despite this hemodilution, there is usually minimal change in mean corpuscular volume (MCV) or mean corpuscular hemoglobin (Hb) concentration (MCHC). The increase in iron demand during pregnancy is met by increased iron absorption. The maternal plasma erythropoietin level increases during pregnancy and reaches its peak in the third trimester [5, 7–9]. This accelerates erythropoiesis, but hemoglobin concentration and hematocrit decrease.

3. Iron metabolism

In adult men, there is usually little iron loss from the body. Because females lose iron during menses, their iron needs are greater. Usually only around 4% of the ingested iron is absorbed in the upper part of the small intestine, mainly in the ferrous state, while the majority is ingested in the ferric state. Many metal-binding proteins bind not only to iron but the other metals such as zinc and copper. After crossing the intestinal cells, most of the absorbed iron is bound to apoferritin forming ferritin. Usually around 35% of the transferritin is saturated with iron. The details of the process, involving intestine, plasma, liver, and bone marrow are shown in the **Figure 1**.

http://dx.doi.org/10.5772/63211

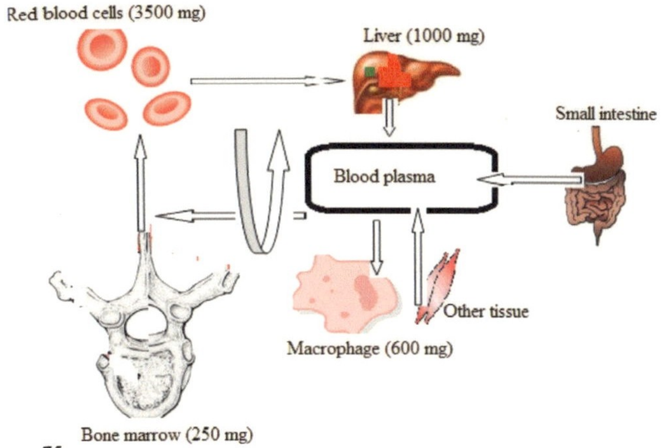

75

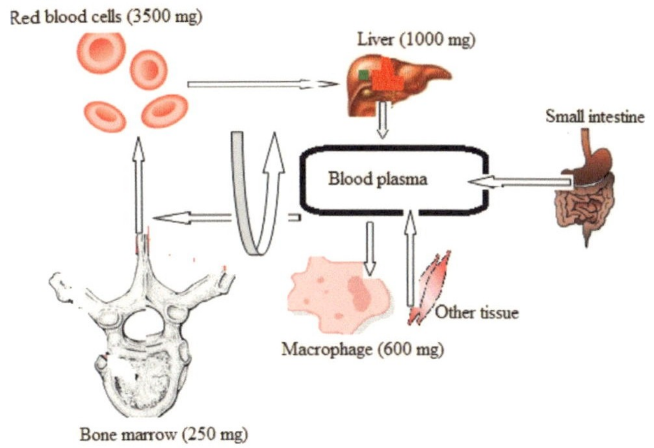

Figure 1. Iron metabolism

4. Iron requirements during pregnancy

The daily requirement of iron is around 1.5 mg in nonpregnant women. This requirement increases dramatically during pregnancy to reach 6–7 mg/day (total 1000 mg) with advanced gestational age. Pregnancy causes a twofold to threefold increase in the requirement for iron and a 10- to 20-fold increase in folate requirement. The increase in demand for iron is mainly due to fetal requirement, placenta, blood volume, tissue accretion, and the intra-partum potential for blood loss [5, 10]. This sixfold increase in need is quite difficult to meet with diet alone, especially in situations of poverty. In many underdeveloped countries, pregnant woman may have depleted iron stores and/or iron-deficiency anemia and, therefore, are at increased risk of becoming anemic during pregnancy and developing the adverse consequences of iron deficiency and anemia. For this reason iron supplementation during pregnancy is very important to keep the maternal hemoglobin within the normal range. It was previously thought that even in the absence of sufficient iron supplementation, fetal hemoglobin production was not impaired because the fetus obtains iron even if the mother suffers from severe anemia. This is now an obsolete theory, and maternal anemia may lead to fetal anemia and many other perinatal adverse effects.

5. Folate during pregnancy

The normal level of folic acid is not sufficient to prevent megaloblastic changes in bone marrow in about 25% of pregnant women. Moreover, folic acid deficiency is more likely to occur in twin pregnancy, and in women taking anti-convulsion and sulfa-containing drugs. All pregnant women in developing countries should receive daily supplementation of 60 mg iron and 40 mg folic acid. Folate level is affected by sickle cell disease, malaria, and hemolytic anemia. The issue of folate deficiency has received global attention due to its association with neural tube defects.

6. Definition of anemia in pregnancy

The World Health Organization (WHO) defines anemia during pregnancy as a hemoglobin concentration <11 g/dl. However, this cutoff value for hemoglobin concentration is affected by many factors such as ethnicity, altitude, and smoking [10]. Anemia in pregnancy is further classified as mild/moderate (Hb 7–10.9 g/dl) and severe (Hb < 7 g/dl) [2]. The Centers for Disease Control and Prevention (1990) defined anemia as hemoglobin <11 g/dl in the first and second trimesters and <10.5 g/dl in the third trimester. This is based on the reduction in hemoglobin level during pregnancy caused by the disproportion in volume expansion between the plasma and erythrocytes. This disproportion is considerably greater during the second trimester. Postpartum anemia is defined by the WHO as hemoglobin <10 g/dl [10].

7. Etiology of anemia

There are several different factors responsible for anemia. The most common is iron deficiency anemia (IDA), which is generally assumed to represent 50% of cases [11]. Among the various risk factors for IDA nutritional or low iron intake together with acute blood loss are the leading causes. During pregnancy, symptoms such as nausea and vomiting together with other contributing factors may cause maternal anemia; the other factors include history of heavy menstruation, high parity, short birth spacing, lack of antenatal nutritional education, and multiple pregnancy. Malabsorption interferes with iron absorption and parasitic infestation such as hookworm may also lead to low hemoglobin levels. Iron absorption is enhanced by ascorbic acid and inhibited by phytic acid and tannins present in tea, coffee, and chocolate.

The second common leading cause of anemia in pregnancy is folic acid deficiency. Other micronutrient deficiency such as vitamin A, B12, and riboflavin, zinc, and copper may also contribute to anemia. Malaria, hookworm infestation, infection, and deficiency of a number of micronutrients are leading causes of anemia during pregnancy. The relative contribution of each of these factors to anemia during pregnancy varies greatly by geographical location. Iron deficiency in anemic subjects in poor communities may be complicated by one or more additional micronutrient deficiencies. The etiologic pattern of anemia during pregnancy is often complex such that, for example, infection and nutritional deficiencies coexist.

Obstetric/gynecologic
Previous history of menorrhagia/metrorrhagia
History of miscarriage
Fibroid
Multiple pregnancies
Teenagers
Infections
Infections, for example, urinary tract infection
Parasitic infections, for example
Malaria, schistosomiasis, and hookworms
HIV
Helicobacter pylori
Bleeding from other site
Peptic ulcer
Hemorrhoids
General
Pica, for example, eating mud
Nutrition habits, for example, vegetarian

Table 1. Causes of and predisposing factors for anemia during pregnancy.

Other etiologies for anemia in pregnancy include malaria, chronic infection including HIV/AIDS, hemolytic anemia, thalassemia, and sickle cell disease.

Pregnancy is suggested as a possible cause for aplastic anemia due to the suppression of hematopoiesis by placental lactogens [12]. This is supported by the clinical observation that pregnancy-associated aplastic anemia is frequently self-limiting, ending with delivery. Pregnancy is one cause of bone marrow suppression, and aplastic anemia is likely to be immune-mediated since pregnancy is a state of hypo-immunity, likely involving suppression by cytotoxic T lymphocytes. In patients with aplastic anemia, CD4 and HLA-DR+ are detectable in both blood and bone marrow. The cells produce inhibitory cytokines such as tumor necrosis factor and gamma interferon, which affect the mitotic cells and induce nitric oxide synthase and nitric oxide production by bone marrow cells, related to immune-mediated cytotoxicity and elimination of hematopoietic cells.

Table 1. shows causes and predisposing factors for anemia in pregnancy.

8. Diagnosis

While mild anemia is usually asymptomatic and may be detected during routine prenatal check up for hemoglobin, moderate and severe anemia may present with different symptoms, including fatigue, dizziness, tiredness, lethargy, fainting, palpitation, symptoms of congestive heart failure, and leg swelling. In severe cases, there may be difficulty in swallowing and/or blindness if there is a vitamin A deficiency. It is worth mentioning that some of these symptoms can overlap and hence be attributed to symptoms detected in normal pregnancy (**Table 2**).

Fatigue
Dizziness
Tiredness
Lethargy
Fainting
Palpitation
Symptoms of congestive heart failure
Legs swelling

Table 2. Symptoms of anemia during pregnancy.

Pallor and physical findings of iron deficiency may also be present, such as angular stomatitis, smooth tongue, and koilonychias, in the **Figure 2**.

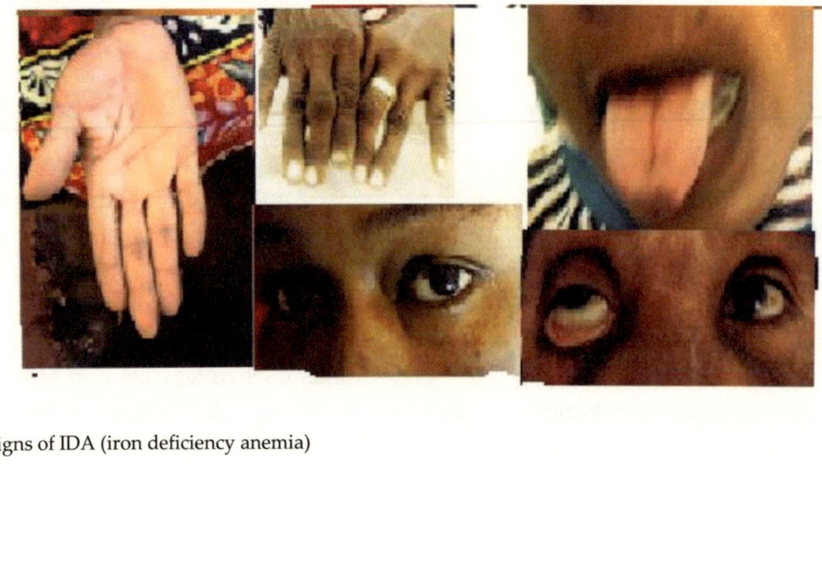

Figure 2. Signs of IDA (iron deficiency anemia)

http://dx.doi.org/10.5772/63211

8.1. Signs of IDA

An abdominal examination to rule out enlarged spleen and/or liver is mandatory in approaching anemic pregnant woman. A complete blood picture (include peripheral blood film) is the first step in tailoring the next investigations aimed at determining etiology as shown in the **Figure 3**.

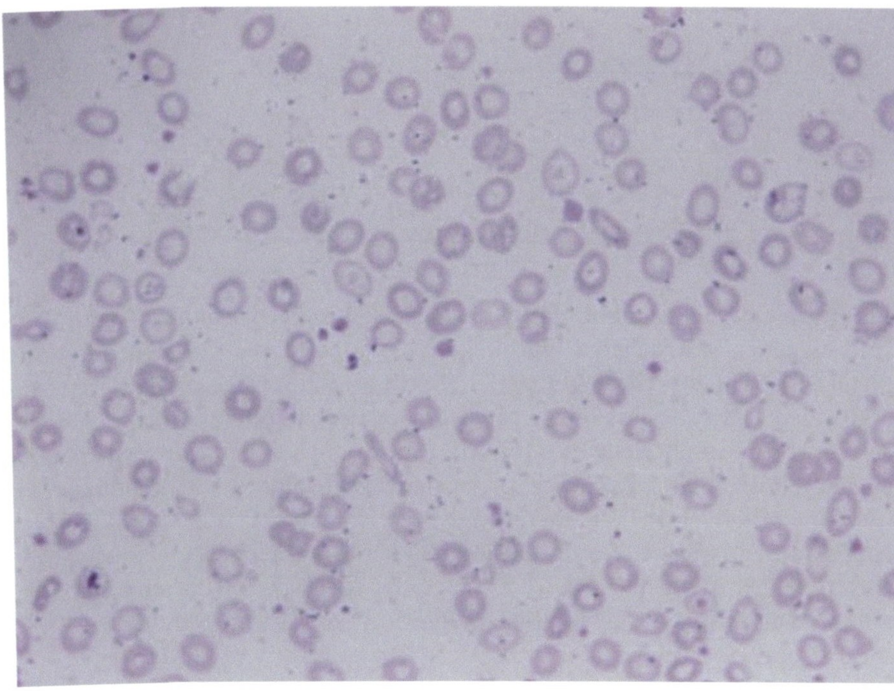

Figure 3. Peripheral blood picture in IDA

These may include stool examination for hookworms, hemoglobin electrophoresis, and tests for infectious organisms such as malaria, tuberculosis, and HIV. Bone marrow aspirate or biopsy may be needed to diagnose the underlying cause of anemia.

The definition and identification of iron deficiency have been problematic, especially in situations in which chronic inflammation is present. The gold standard for identifying iron deficiency anemia has been the examination of suitably stained bone marrow aspirates for storage iron as hemosiderin. Biochemical measurement of iron status is influenced by inflammation and clearly defined and validated cutoffs for diagnosing iron deficiency in pregnancy in the presence of coexisting infection have been lacking. A lowered MCV is the most sensitive indicator of iron deficiency, where serum iron is low and the total binding capacity raised. Macrocytosis with megaloblastic changes in bone marrow in an indicator of folate deficiency anemia (**Table 3**).

Factor	Lower cutoff
Hemoglobin	11 g/dl
Hematocrit	30%
Mean corpuscular volume	80 fl
Mean cell hemoglobin	28 pg
Mean cell hemoglobin concentration	32 g/dl
Serum ferritin	12 μg/l
Total iron binding capacity	15%

Table 3. The lower cut off points of the hematological indices during pregnancy.

9. Consequences of anemia in pregnancy

Anemia during pregnancy is associated with increased maternal morbidity and mortality. Anemia in pregnancy is associated with negative consequences for both the woman and neonate. Fetal anemia, low birth weight, preterm birth, and stillbirth have been associated with anemia [13]. Anemia was observed as a predictor for poor perinatal outcomes such as fetal anemia and low birth weight deliveries [13, 14]. A meta-analysis showed that anemia during early pregnancy, but not late pregnancy, is associated with slightly increased risk of preterm delivery and low birth weight [15]. Interestingly, recent reports revealed that the prevalence of preeclampsia and eclampsia was significantly higher in women with severe anemia [16]. In some African countries, anemia was reported to be associated with stillbirth [17, 18]. In addition, there is also association between anemia and postpartum hemorrhage and pulmonary edema.

Maternal
Anemic heart failure
Maternal death
Septicemia
Preterm labor
Low tolerance to systemic diseases, for example, heart disease
Postpartum hemorrhage
Lack of tolerance of minimal bleeding
Impaired ability to push during the second stage of labor
Impaired lactation
Breast milk of low nutritional value
Deep venous thrombosis

http://dx.doi.org/10.5772/63211

Puerperal psychosis
Cognitive abnormalities
Perinatal
Intrauterine growth restriction
Intrauterine fetal death
Low birth weight
Fetal anemia
Low APGAR score
Increased perinatal motility
Increased infant death

Table 4. Complications/adverse effects of anemia during pregnancy.

10. Treatment and prevention

10.1. Dietary advice

As physiological iron requirements are several times higher in pregnancy than they are in the nonpregnant women, the recommended daily intake of iron for the second half of pregnancy is 30 mg with iron absorption increasing threefold. The amount of iron absorbed depends upon the following factors: (1) amount of iron in the diet, (2) its bioavailability, and (3) physiological requirements. Dietary heme iron is found mainly in red meats, fish, and poultry. Heme iron absorption is twofold to threefold greater than non-heme iron. Moreover, meat contains organic compounds (including peptides), which promote the absorption of iron from other less bioavailable non-heme iron sources. While heme iron is more readily absorbed than non-heme iron, the latter still forms approximately 95% of dietary iron intake. Ascorbic acid significantly increases iron absorption from non-heme sources, with the magnitude of this effect concordant with the increase in quantity of vitamin C in the meal. The bioavailability of non-heme iron is enhanced by germination and fermentation of cereals and legumes which results in a decrease in the phytate content, a food constituent that hinders iron absorption. Tannins in tea and coffee hinder iron absorption on consumption with or shortly after a meal.

Nutritional education is the main objective of antenatal care to assist in the prevention of anemia. In addition, family planning and control of birth spacing is another preventive measure that should be considered by health-care providers. The WHO jointly with the International Nutritional Anemia Consultative Group and the United Nations Children's Fund recommend routine supplements of 60 mg iron per day and 400 μg folate per day to all pregnant women for at least 6 months. This guideline also recommends continuation until 3 months postpartum in areas of high prevalence of anemia (>40%). The standard oral preparation, Fefol, comprising 100 mg iron and 350 μg folate, is suitable for both prevention and treatment. Parenteral iron does not provide rapid correction of hemoglobin levels compared with oral form but is an option for those with poor compliance and who cannot tolerate the oral formulation. It is also suitable in cases of malabsorption. The maximum rise in hemoglobin achievable with either oral or parenteral formulations is 0.8 g/dl/week. Blood transfusion is indicated in cases of severe anemia (Hb% < 7 g/dl) and anemia in late pregnancy when delivery is due.

Referral to secondary care level should be considered if any of the following situations exist:

1. Significant symptoms and/or severe anemia (Hb < 70 g/l), or
2. Advanced gestation (>34 weeks), or
3. If there is no rise in Hb at 2 weeks.
4. In non-anemic women who are at increased risk of iron depletion such as those with:
   a. Previous anemia,
   b. Multiple pregnancy,
   c. Consecutive pregnancies with <1 year's interval between
   d. Vegetarians
   e. Women at high risk of bleeding
   f. Pregnant teenagers
   g. Jehovah's witnesses.

10.2. Postnatal anemia

The WHO definition for postnatal anemia is Hb < 10 g/dl. Complete blood count should be checked within 48-hour post-delivery in all women with an estimated blood loss >500 ml and in women with uncorrected anemia in the prenatal period or symptoms suggestive of postpartum anemia. Elemental iron 100–200 mg daily for at least 3 months should be offered to women with Hb < 100 g/l, who are hemodynamically stable, asymptomatic, or have mild symptoms, and a repeat complete blood count and ferritin level assessment should be undertaken to ensure hemoglobin and iron stores are replete [17].

10.3. Parenteral iron therapy

Indications for parenteral iron therapy [19]:

1. absolute noncompliance with oral iron therapy
2. intolerance to oral iron therapy
3. proven malabsorption.

Parenteral iron therapy bypasses the natural gastrointestinal regulatory mechanisms to supply nonprotein-bound iron to the red blood cells. It is characterized by fast increases in Hb and

http://dx.doi.org/10.5772/63211

better replenishment of iron stores compared with oral therapy, particular iron sucrose. However, issues concerning its safety are waiting to be addressed.

Contraindications for parenteral iron use are as follows:

1. history of anaphylaxis or reactions to parenteral iron therapy

2. first trimester of pregnancy

3. active acute or chronic infection

4. chronic liver disease.

Appropriate setting and staff trained in management of anaphylaxis should be available on contemplating usage of parenteral iron.

10.4. Dealing with delivery of women with iron deficiency anemia

With good practice, this situation should generally be avoided; nonetheless, there are instances when women book late have recently arrived from abroad or have not engaged with antenatal care. In such circumstances, it may be essential to take active measures to minimize blood loss at parturition. Attention should be paid to delivery in hospital, securing an intravenous access and blood group and save, and consideration of active management of the third stage of labor to reduce postpartum blood loss.

10.5. Blood transfusion: indications and risks

There are multiple potential hazards from blood transfusions but most arise from clinical and laboratory errors. Moreover, specific risks for women of child-bearing age include the potential for transfusion-induced sensitization to red blood cell antigens, creating a future risk of fetal hemolytic disease. Massive obstetric hemorrhage is widely appreciated as an important cause of morbidity and mortality and necessitates prompt use of blood and components as part of appropriate management.

Both clinical assessment and hemoglobin concentration are of immense significance postpartum to decide on the optimum method of iron replacement. In the absence of bleeding, the decision to transfuse blood should be made on an informed individual basis.

Blood transfusion should be reserved for women with:

a. continued bleeding or at risk of further bleeding,

b. imminent cardiac compromise

c. significant symptoms requiring urgent correction.

If, after careful consideration, elective blood transfusion is needed, women should be fully counseled about potential risks and given written information, and consent should be obtained.

10.6. Prophylaxis

Efforts aimed at preventing iron deficiency and iron deficiency anemia among pregnant women include iron supplementation, fortification of staple foods with iron, increasing health and nutritional awareness, combating parasitic infections, and improvement in sanitation [20]. A prophylactic dose of 300 μg (0.3 mg) daily during pregnancy was proposed in 1968 by the WHO.

During pregnancy, women need iron supplementation to ensure they have sufficient iron stores to prevent iron deficiency [21]. Hence, in most developing countries, iron supplements are used extensively during pregnancy to prevent and correct iron deficiency and anemia during gestation.

A dose of 60 mg of elemental iron was accepted as standard supplemental dose in 1959, depending on estimates of iron needs during pregnancy [22]. This has since been endorsed by several experts [23, 24]. Gastrointestinal discomfort is a common observation among women consuming large amounts of supplemental iron, especially if taken on an empty stomach. Gastrointestinal side effects are recognized as the critical adverse effect on which the tolerable upper limit of intake for iron is determined. Use of high-dose iron supplements commonly leads to gastrointestinal manifestations, such as constipation nausea, vomiting, and diarrhea, with the frequency and severity being determined by the amount of elemental iron released in the stomach.

10.7. Folic acid supplementation

Following publication of a number of studies supporting the periconceptional use of folic acid in the prevention of neural tube defects, the supplemental dose was increased to 400 μg (0.4 mg) of folic acid daily in 1998. This dose was considered to provide more folic acid than needed to produce an optimal hemoglobin response in pregnant women. If supplementation is delayed till after the first trimester of pregnancy, it will not contribute to preventing birth defects [25]. Interestingly, a recent Cochrane review showed that supplement with folic acid, alone or in combination with vitamins and minerals, prevents neural tube defects, but it does not have a clear effect on other birth defects [26].

Likewise, it has been found that folic acid alone, or in combination with vitamin and mineral supplements during pregnancy, improved iron status in women without affecting perinatal anemia, perinatal mortality or other infant outcomes [27, 28].

11. Hemoglobinopathies

Hemoglobinopathies such as thalassemias and sickle cell diseases should be considered during pregnancy because of their impact on maternal and perinatal outcomes. They are genetic disorders of hemoglobin structure and synthesis and may transmit to the offspring. The main clinical manifestation of these disorders during pregnancy is anemia. Usually the iron store is quite normal necessitating folate supplementation without iron to avoid iron overload. Preconceptual counseling is a very important issue in patients with hemoglobinopathies. It allows establishment of the hemoglobin status of the parents and prediction of the likelihood of an affected offspring [29].

11.1. Sickle cell disease

Sickle cell disease is caused by the substitution of glutamic acid by valine at position 6 of the globin chain. It includes sickle cell anemia, sickle hemoglobin C disease, sickle beta thalassemias, and sickle cell anemia with alpha thalassemia. Sickling and crystallization of the hemoglobin are induced by de-oxygenated states such hypoxia, acidosis, and dehydration. Almost always the patients are already diagnosed prior to pregnancy, but the diagnosis is made by hemoglobin electrophoresis. Sickle cell disease substantially increases maternal and perinatal mortality. It leads to miscarriage, intrauterine growth restriction, preterm labor, preeclampsia, abruptio placentae, and thrombosis [29]. There is also an increased incidence of infection, and the sickle cell crisis should be managed as aggressively as in nonpregnant women. The management of sickle cell disease in pregnancy should be in collaboration with hematologist. Folic acid supplementation with avoidance of iron is very important together with penicillin prophylaxis. The patient should be removed from factors that may have triggered the crisis. These may include dehydration in early pregnancy (hyperemesis gravidarum) and during labor. Regular antenatal monitoring, serial growth scans, and intrapartum avoidance of dehydration, hypoxia, acidosis, and infection are very important, as are consideration of analgesia and anesthesia. The routine use of prophylactic blood transfusion is controversial in this situation.

11.2. Thalassemias

The genetically determined hemoglobinopathies termed thalassemias are characterized by impaired production of one or more of the normal globin peptide chains. Thalassemias occur according to which globin chain is deficient.

Alpha thalassemia minor (three normal alpha gene) is usually asymptomatic, but the patient may become anemic. Alpha thalassemia major is incompatible with life, and the fetus is severely hydropic. Beta thalassemia minor may also be a symptomatic but may present with iron deficiency anemia with lowered MCV, MCH, and MCHC. The patient will need oral folate and iron supplementation. Beta thalassemia major in adults presents with iron overload.

12. Acquired hemolytic anemia

This is an uncommon type of anemia and is either primary or secondary. It is usually due to antibody production. Secondary hemolytic anemia may be due to chronic infection, drugs, or connective tissue disease. Typically both direct and indirect Coombs tests are positive and spherocytosis and reticulocytosis are the typical characteristics of a peripheral blood smear. Steroids are usually effective treatment (prednisolone 1 mg/kg/day). The presentation and symptoms depend on the severity of hemolysis. Very rarely, as in gestational thrombocytopenia, there is pregnancy-induced hemolytic anemia. However, usually the condition is benign and resolves spontaneously. Some obstetric conditions such as pre-eclampsia and eclampsia may induce micro-angiopathic hemolysis, and this might progressed to hemolysis, elevated liver enzymes, and low platelets (HELLP) syndrome [29].

Author details

Ishag Adam[1*] and Abdelaziem A. Ali[2]

*Address all correspondence to: ishagadam@hotmail.com

1 Department of Obstetrics and Gynecology, Faculty of Medicine, University of Khartoum, Khartoum, Sudan

2 Faculty of Medicine, Kassala University, Khartoum, Sudan

References

[1] WHO/CDC. *Worldwide prevalence of anemia 1993–2005. WHO Global Database on Anemia*. Geneva, World Health Organization. 2008. http://whqlibdoc.who.int/publications/2008/9789241596657_eng.pdf. Accessed 1 December 2012.

[2] WHO/UNICEF/UNU. *Iron deficiency anaemia assessment, prevention, and control: a guide for programme managers*. Geneva, World Health Organization. 2001. http://www.who.int/nutrition/publications/en/ida_assessment_prevention_control.pdf. Accessed December 2012.

[3] Allen L.H. Anemia and iron deficiency: effects on pregnancy outcome. *Am J Clin Nutr*. 2000;71(5 Suppl):1280S–1284S.

[4] Breymann C. Iron deficiency anemia in pregnancy. *Semin Hematol*. 2015;52(4):339–47.

[5] Koller O. The clinical significance of hemodilution during pregnancy. *Obstet Gynecol Surv*. 1982;37:649–652.

[6] Murphy JF, O'Riordan J, Newcombe RG, Coles EC, Pearson JF. Relation of haemoglobin levels in first and second trimesters to outcome. *Lancet*. 1986;3:992–995.

[7] Hunter S, Robson SC. Adaptation of the maternal heart in pregnancy. *Br Heart J*. 1992;68:540–543.

[8] Mahendru AA, Everett TR, McEniery CM, Wilkinson IB, Lees CC. The feasibility of prospectively studying maternal cardiovascular changes from before conception. *Hypertens Res*. 2013;36(8):698–704.

[9] Adamson SL, Lu Y, Whiteley KJ, Holmyard D, Hemberger M, Pfarrer C, Cross JC. Interactions between trophoblast cells and the maternal and fetal circulation in the mouse placenta. *Dev Biol*. 2002;250:358–373.

[10] World Health Organization. 1990. The prevalence of anemia in women: a tabulation of available information. Geneva: Maternal Health and Safe Motherhood Programme, World Health Organization. p. 100.

[11] Van den Broek NR, Letsky EA. Etiology of anemia in pregnancy in south Malawi. *Am J Clin Nutr*. 2000;72(1 Suppl):247S–256S.

[12] *Iron deficiency anemia: assessment, prevention, and control. A guide for programme managers*. Geneva, World Health Organization, 2001 (WHO/NHD/01.3).

[13] Elhassan EM, Abbaker AO, Haggaz AD, Abubaker MS, Adam I. Anemia and low birth weight in Medani, Hospital Sudan. *BMC Res Notes*. 2010;28(3):181.

[14] Kidanto HL, Mogren I, Lindmark G, Massawe S, Nystrom L. Risks for preterm delivery and low birth weight are independently increased by severity of maternal anemia. *S Afr Med J*. 2009;99(2):98–102.

[15] Bondevik GT, Lie RT, Ulstein M, Kvale G. Maternal hematological status and risk of low birth weight and preterm delivery in Nepal. *Acta Obstet Gynecol Scand*. 2001;80(5): 402–408.

[16] Ali AA, Rayis DA, Abdallah TM, Elbashir MI, Adam I. Severe anemia is associated with a higher risk for preeclampsia and poor perinatal outcomes in Kassala hospital, eastern sudan. *BMC Res Notes*. 2011;4:31.

[17] Murphy JF, O'Riordan J, Newcombe RG, Coles EC, Pearson JF. Relation of hemoglobin levels in first and second trimesters to outcome of pregnancy. *Lancet*. 1986;1(8488): 992–5.

[18] Ali A A, Adam I. Anemia and stillbirth in Kassala hospital, eastern Sudan. *J Trop Pediatr*. 2011;57(1):62–4.

[19] Pavord S, Myers B, Robinson S, Allard S, Strong J, Oppenheimer C. British Committee for Standards in Haematology. UK guidelines on the management of iron deficiency in pregnancy. *Br J Haematol*. 2012;156(5):588–600

[20] Royal College of Obstetricians and Gynaecologists. 2007. Blood transfusions in obstetrics. RCOG Green-top guideline.

[21] Chaparro C. Essential delivery care practices for maternal and newborn health and nutrition. Informational Bulletin. Washington, DC, Pan American Health Organization, 2007:1–4. http://www.paho.org/english/ad/fch/ca/ca_delivery_care_practices_eng.pdf. Accessed 1 December 2012.

[22] Bothwell TH. Iron requirements in pregnancy and strategies to meet them. Am J Clin Nutr. 2000;72(Suppl. 1):S257–S264.

[23] Iron deficiency anaemias: Report of a WHO study group. Geneva, World Health Organization. 1959. WHO Technical Report Series, No. 182. http://whqlibdoc.who.int/trs/WHO_TRS_182.pdf. Accessed 1 December 2012.

[24] Nutritional anaemias: Report of a WHO scientific group. Geneva, World Health Organization 1968. WHO Technical Report Series, No. 405; http://whqlibdoc.who.int/trs/WHO_TRS_405.pdf. Accessed 1 December 2012.

[25] Stoltzfus R, Dreyfuss M. Guidelines for the use of iron supplements to prevent and treat iron deficiency anaemia. Washington, DC, ILSI Press, 1998 (http://www.who.int/nutrition/publications/micronutrients/guidelines_for_Iron_supplementation.pdf. Accessed 1 December 2012).

[26] De-Regil LM, Peña-Rosas JP, Fernández-Gaxiola AC, Rayco-Solon P. Effects and safety of periconceptional oral folate supplementation for preventing birth defects. *Cochrane Database Syst Rev*. 2015;12:CD007950.

[27] Mei Z, Serdula MK, Liu JM, Flores-Ayala RC, Wang L, Ye R, Grummer-Strawn LM. Iron-containing micronutrient supplementation of Chinese women with no or mild anemia during pregnancy improved iron status but did not affect perinatal anemia. *J Nutr*. 2014;144(6):943–8.

[28] Liu JM, Mei Z, Ye R, Serdula MK, Ren A, Cogswell ME. Micronutrient supplementation and pregnancy outcomes: double-blind randomized controlled trial in China. *JAMA Intern Med*. 2013;173(4):276–82

[29] Naik RP, Lanzkron S. Baby on board: what you need to know about pregnancy in the hemoglobinopathies. *Hematol Am Soc Hematol Educ Program*. 2012;2012:208–14.

Chapter 7

Malaria, Schistosomiasis, and Related Anemia

Gasim I Gasim and Ishag Adam

Additional information is available at the end of the chapter

http://dx.doi.org/10.5772/63396

Abstract

Parasitic infections (e.g., malaria and helminthiases) have a huge impact on public health in endemic areas. Moreover, parasitic infestations are prominent causes of anemia in the tropics and subtropics, further perpetuated by malnutrition, inflammatory, and genetic diseases. Anemia-associating parasitic infections vary depending on the requirements and pathophysiology of the parasites. There is an interplay between different factors that can be segregated as host and parasite factors, resulting in severe anemia accompanying these parasitic infestations. The pathophysiological mechanisms leading to anemia associated with the different parasites vary greatly, including hemolysis, anemia of inflammation, bone marrow suppression, and micronutrients deficiency. The major means to deal with this anemia include prevention and treatment of such infestations.

Keywords: malaria, schistosomiasis, anemia, pathogenesis, parasite

1. Overview

Parasitic infestations (e.g., malaria and helminthiases) have an enormous impact on public health in endemic areas. Moreover, parasitic infections are leading causes of anemia in the tropics and subtropics, worsened by malnutrition, inflammatory, and genetic diseases. Anemia-associating parasitic infections vary depending on the requirements and pathophysiology of the parasites. It sounds reasonable that the closer the parasite's association with the red blood cells (RBCs), the more severe the expected anemia. On speaking about blood parasites, malaria is the most important and well-known infection worldwide. Anemia is a clinical condition where the values of hemoglobin, hematocrit, or RBCs counts are more than two standard deviations below the mean for a particular age and sex, with severe anemia characterized by hemoglobin of less than 5 g/dL. Anemia develops as a consequence of blood loss, when red cells are destructed prematurely, or when the normal erythroid production of red cells is disturbed. These mechanisms often overlap with a number of factors contributing to anemia. Among the important causes of increased cell destruction leading to acquired hemolytic anemia is malaria. Hypersplenism and splenomegaly as in hyper-reactive malaria also play an important role in hemolysis. Another blood parasite of importance is schistosomiasis which is caused by a blood fluke that undergoes a complex life cycle using a species of freshwater snail. Adult flukes pair post maturation inside a human host, for life and begets thousands of eggs that brings harm to organs and are excreted in urine and feces. The larvae hatching from the eggs manage their way into the snails that in turn begets vast numbers of larvae capable of penetrating the human skin. The fluke lives in the veins, urinary bladder, and large intestine of their human hosts and borrow molecules from their hosts to put on their surfaces so that the hosts' immune system would not recognize them as strange.

2. Malaria and anemia

Malaria is an ancient febrile illness that continues to jeopardize human existence. It is one of the major killers, particularly among the tropical countries in Africa, Southeast Asia, and Latin America which is a mosquito-borne disease the characteristic symptoms of which are cyclical bouts of fever with muscle stiffness, shivering, and sweating whose periodicity reflects the intraerythrocytic cycle. Malaria is a disease resulting from the parasitic infestation by *Plasmodium* species, such as *Plasmodium falciparum, P. malariae, P. ovale, P. vivax,* and *P. knowlesi* with *P. falciparum* being the most virulent. Malaria is estimated to be a burden for over 200 million people, leading to more than one million fatalities annually. The main vector for this *Plasmodium* is *Anopheline* species, which are most common tropical inhabitants [1]. Malaria is dependent on the vector-human cycle, and it affects impoverished people in the suburbanized endemic areas with economic and social consequences. Despite decades of efforts on the battle against malaria, it remains to be an important health threat in tropical areas [2]. Malaria can manifest a vast clinical spectrum from silent carrier to fatal shock.

3. Common clinical features of malaria

Fever

Chills

Headache

Myalgia

Malaise

Anemia

Petechie

http://dx.doi.org/10.5772/63396

4. Manifestations of severe disease

Seizures

Jaundice

Mental confusion

Renal failure

Acute respiratory disease syndrome (ARDS)

Coma

Thrombocytopenia

Severe anemia

Hypoglycemia

Hyperparasitemia

Hypotension

Bleeding

Blackwater fever

5. Genetic basis of malaria-associated anemia

Malaria is a polygenic disease, and the genetic basis of malaria-related anemia is under study. Variable genes have been shown to be involved in host predisposition to the severe forms of malaria, part of which is malaria-related anemia; nevertheless, it is likely that there are undetected malaria-susceptibility genes. It has been found that severe malaria-related anemia is associated with a number of genes, such as FcγRIIA-131H/FcγRIIIB-NA2 haplotype, interleukin-13 promoter polymorphisms (-7402 T/G and -4729G/A), and TNF-238 A allele [3–5]. The host-parasite interaction is complex and not fully understood. Such an interaction leads to a release of a number of cytokines, resulting in the so-called "cytokine storm" in the setting of severe malaria, where injurious cytokines and small molecules become dysregulated and results in a systemic inflammatory response syndrome (SIRS)-like state characterized by high circulating levels of tumor necrosis factor (TNF) and nitric oxide. However, evidence of direct correlation between severe malaria and the activity of these markers is limited [6]. Elevated serum levels of the different cytokines such as TNF, lymphotoxin, interleukins 6, 10, 12, and 18, and macrophage inflammatory protein (MIP)-1 are seen in the setting of malaria. Nevertheless, more studies are needed to clarify whether these predate or follow clinical markers of severe infection [6]. It is proposed that interferon-regulated gene transcripts influence the inflammatory response to cytokines, and these results demonstrated previously undiscovered transcriptional changes in the host that might govern the development of malaria-associated syndromes, such as anemia and metabolic dysregulation [7]. On the other hand, a number of genes were found to be protecting against malarial anemia such as SCGF, also called C-type lectin domain family member 11A [CLEC11A]), IL12Bpro-2/3′ UTR-T haplotype, FcγRIIA-131H/FcγRIIIB-NA1 haplotype, and NOS2 promoter polymorphisms, along with HLA class II allele DQB1*0501 [3, 8, 9]. In addition, specific genes for commonly inherited diseases found in the tropics are also known for their role in resistance to malaria-related anemia. Such effects imposed by these genes are thought to reflect good examples in the natural selection process in the tropical area. Upon discussing the genetic basis of anemia, it is prudent to speak about the different hemoglobinopathies and their genes such as sickle cell anemia. The most commonly mentioned of such genes are Hb S, hemoglobin E, glucose-6-phosphate dehydrogenase deficiency, pyruvate kinase deficiency hereditary elliptocytosis (HE), and thalassemia genes where several studies have found an inhibitory effect of thalassemic gene on malaria-related anemia [10].

6. Pathophysiology of malaria-associated anemia

Anemia is one of the primary pathophysiological events contributing to fatal malaria [11]. Severe and refractory anemia causes hypoxia and leads to heart failure in malaria patients [12]. A number of mechanisms contribute to the pathogenesis of malaria-related anemia, such as erythrocyte destruction and phagocytosis, sequestration of infected RBCs, dyserythropoiesis, and bone marrow suppression. Erythrocyte lysis could be due to hemolysis of either parasitized red cells or non-parasitized cells. Red cells of malaria patients suffering from severe anemia have been found to display abnormal distribution of the different membrane phospholipids, for example, (phosphatidylserine (PS), phosphatidylcholine, and phosphatidyl ethanolamine) non-parasitized, along with membrane damage induced by heme released from the digestion of hemoglobin by the parasite which underwent lipid peroxidation [6]. Expression of specific antibodies directed against the variant parasite antigens (PfEMP-1) surface of the red cells that results in opsonization of the infected red cells [13]. Interestingly, lysis of cells is not confined to parasitized RBCs only where it has been found that non-parasitized erythrocytes inside the parasite culture showed a significant increase in the lipid peroxide genesis and vulnerability to lysis [14]. Moreover, a direct correlation between membrane lipid peroxidation and peroxide hemolysis exists, both before and after monocyte exposure, implying a primary role of membrane peroxidation in red cell lysis. Children with malaria showed low levels of the antioxidant α-tocopherol in the membrane of red cells, a finding that might support the hypothesis that local antioxidant consumption may contribute to erythrocyte loss. It is also suggested that parasite products forming part of the immunoglobulin-antigen complexes retained on non-parasitized erythrocytes include the *P. falciparum* ring surface protein 2 (RSP-2), which results in opsonization of these non-parasitized RBCs and thus provides a mechanism of removing non-parasitized RBCs [6]. A steady decline in the hemoglobin level accompanied by an inappropriate reticulocyte response occurs following an acute malarial infection, where this sort of anemia is explained by sequestration of iron in the spleen and other reticuloendothelial system organs together with a shortened red cell survival. It is considered to be very rare in malaria despite the presence of some

http://dx.doi.org/10.5772/63396

evidences in support of the nature of the normocytic normochromic anemia with evidence of malaria-related anemia due to hemolysis; however, recent data indicate that these mechanisms (singly or in combination) do not fully explain the severity of this anemia. Dyserythropoiesis is proposed to play a role in malaria-related anemia, although malaria-related anemia might partially be attributed to sequestration of parasitized cells, the continuous reduction in hemoglobin level for several weeks after the acute episode should raise the possibility of involvement of other factors. Hematologic studies have shown that bone marrow suppression and inefficient erythropoiesis have an important share in the severe anemia of malaria infection [15]. Host mechanisms in control of suppression of erythropoiesis might involve an exaggerated and steady innate immune response or a pathologic alteration of the T-cell differentiation response along with the concomitant production of certain proinflammatory cytokines. We are not to over-look the erythrocytes destructed by the spleen and reticuloendothelial hyperactivity, where large numbers of both parasitized and non-parasitized red cells are destructed. Dyserythropoiesis and severe anemia attributed to malaria are closely associated with excess release of interferon (IFN)-γ and TNF-α, along with nitrous oxide, which promote enhanced malarial anemia pathogenesis also resulting in bone marrow suppression and erythrophagocytosis. Other cytokines like interleukin (IL)-12 and 18 have also been implicated in dyserythropoiesis. Hemozoin, which is a malarial pigment resulting from incomplete hemoglobin digestion by the parasite, has also been incriminated in the impaired erythroid development through its direct effects on human monocyte function and/or erythroid precursors [16]. Other contributing factors to malarial anemia are coinfection with other organisms such as bacteria, viruses (e.g., human immunodeficiency virus), and helminthiasis. In summary, the pathogenesis of malaria-related anemia seems to be very complex. It is indicated that there is a cardinal defect in erythroid maturation with existence of a significant degree of erythrophagocytosis. Nevertheless, more elaboration on the subject of pathophysiology of malaria-related anemia is needed. Concerning non-falciparum malaria, although not common, but it can be seen in the non-falciparum malarial patients, particularly in cases where hemolysis due to G-6-PD deficiency is encountered and those receiving some drugs inducing hemolysis. The most important drug to be considered on speaking about hemolysis due to G-6-PD deficiency is primaquine, which is an effective antimalarial drug recommended for the dormant hypnozoites of vivax malaria.

7. Mechanisms of anemia in malaria

Increased destruction	Inadequate response to anemia
1. Destruction and lysis of parasitized erythrocytes	Dyserythropoiesis due to: 1. Excess IFN-γ
2. Destruction and lysis of un parasitized erythrocytes	2. Excess TNF-α
3. Drug-induced hemolysis	3. Deficient interleukin-12 production

Increased destruction	Inadequate response to anemia
4. Destruction by the spleen and reticuloendothelial system	**4.** Effect of hemozoin leading to impaired erythroid development

Table 1. Mechanisms of anemia in malaria.

8. Clinical manifestation of malaria-associated anemia

Malaria-related anemia is a frequent manifestation of *P. falciparum* malaria; nevertheless, it is increasingly being reported as a manifestation of *P. vivax* malaria. The most vulnerable groups of people are those under five years old and pregnant women. Furthermore, micronutrient deficiencies caused stunting and also impaired host immunity, thereby increasing the degree to which malaria is associated with low concentrations of hemoglobin, beside increased inflammation, and with increased need for iron in young erythroblasts where the anemia might be severe enough to require blood transfusion. Generally speaking, the spectrum of presentation is broad and influenced by a number of factors that are host related or parasite related or a mixture of both such as the age at presentation, whether it is acute or chronic malaria, the patient's immune status and if he already lives an endemic area or not and the association with other conditions that might worsen or protect against the anemia. Pallor is the most commonly encountered presentation of malarial anemia that can be detected by physical examination and confirmed by a simple hemoglobin test. Additional symptoms requiring referral and blood transfusion are an ejection systolic murmur, change in the consciousness level, the presence of splenomegaly, or malarial parasitemia. In severe malaria-related anemia, it is proposed that cardiac symptoms could be caused by a cardiomyopathy as an after-effect of malarial chronic anemia. Moreover, severe malarial anemia might present with severe lactic acidosis. Severe malarial anemia has been commonly linked to *P. falciparum*; nevertheless, *P. vivax* has been found to cause severe malarial anemia. It is known that parasite density in *P. vivax* malaria plays a significant role that influences the fragility of the RBCs, Heinz body formation, Mean Corpuscular Volume (MCV), Mean Corpuscular Hemoglobin (MCH), and Mean Corpuscular Hemoglobin Concentration (MCHC) levels; there for, the RBCs of the patients recurrently infected with *P. vivax* parasite are imposed to structural and functional dysfunction, finally culminating in anemia. However, the anemia is not an uncommon presentation in the patients with *P. vivax* malaria. This is the case that is commonly seen when antimalarial treatment is used in G-6-PD deficient individuals. As in other types of drug induced hemolytic anemia, drug-induced *P. vivax* malaria hemolytic anemia warrants prompt detection and early management. The type of anemia in *P. falciparum* malaria is that of normocytic and normochromic, and absent reticulocytes. Blackwater fever is another type of hemolytic anemia in malarial patients that tends to present with classical features of hemolysis such as hemoglobinuria. This fever is a special clinical entity that presents with features of acute intravascular hemolysis that classically occurs after the reintroduction of quinine in long-term inhabitants in malaria endemic areas and repeatedly inadequately using it. Those patients suffering from G-6-PD deficiency are at particular risk of this syndrome, when being subject-

http://dx.doi.org/10.5772/63396

ed to oxidant drugs even in the absence of malarial infection. The features of this serious complication are bilious vomiting, prostration, intravascular hemolysis, hemoglobinuria, and renal impairment. This is usually a serious and severe complication that should be avoided.

9. Common clinical presentations of malaria-associated anemia

Pallor

Ejection systolic murmur

Change in the consciousness level in association with splenomegaly and parasitemia

Myocardiopathy

Severe lactic acidosis

Blackwater fever

10. Diagnosis of severe malaria-associated anemia

The severe malarial anemia is defined by the World Health Organization (WHO) as:

A hemoglobin less than 5 g/dL or hematocrit less than 15%.

Parasitemia with more than 100,000 parasites/μL of blood.

Normocytic blood film (thus excluding thalassemia as well as iron, B12, and folate deficiencies).

[17, 6].

11. Management of malaria-associated anemia

The fundamentals of management of malaria-related anemia is based on the main principles of dealing with anemia "improvement of RBC genesis, where decreased RBC production is the fundamental pathophysiology along with RBC replacement and decrease RBC lysis in cases that have increase RBC destruction as the culprit pathophysiology" and fundamentals of management for infection "elimination of the source of infection and control of complications from pathogen virulence, host responses and treatment".

11.1. Role of erythropoietin

Despite the fact that ineffective or inadequate erythropoietin production might contribute to malaria-associated anemia in some settings; nevertheless, studies from endemic areas such as Africa showed that children with malaria have elevated erythropoietin production than expected. Therefore, a plausible explanation is that it is rather the response to erythropoietin which contributes to the pathology rather than synthesis, as seen in the anemia of chronic diseases. And as such, administering erythropoietin is not expected to improve malarial anemia [6]

11.2. Is there any role for blood transfusion?

The blood transfusion for malaria-related anemia is an old practice that has been practiced for a long time, the benefits of which have not been validated. Nevertheless, the use of blood transfusion in management of malaria-related anemia carries a high risk for blood-borne infections, particularly in poor resource settings where screening in the blood bank process is lacking or ineffective.

11.3. Is there any role for iron supplementation?

A group of researchers reported that iron supplementation with antimalarial treatment significantly reduced malaria. Moreover, they refuted the assumption that supplementation during an acute attack of malaria increases the risk for parasitological failure or deaths [18].

11.4. Eliminating source of Infection

Theoretically, the fastest way of getting rid of the source of malarial infection and its products is the blood exchange, where it was thought to decrease the degree of parasitemia, when used as adjunct therapy to quinine; however, since there was no supporting evidence, the CDC is now advising against it [19, 20]. On the other hand, antimalarial drug therapy is considered to be the slower method for getting rid of the source of infection, and is definitely needed to manage malaria-related anemia although some of these drugs are to be used cautiously fearing drug-induced hemolytic anemia.

11.5. Treating coinfecting organisms

Studies addressing the effect of coinfection on malarial anemia showed variable results with complex outcomes on anemia [21]. Similar outcomes were seen with studies dealing with the issue of treating coinfection or not [22].

11.6. Control of complications

It is of paramount importance to bear in mind the early recognition of malarial anemia as one of the serious complications of malaria, and thus it is recommended to include hemoglobin measurement as part of the management plan of malaria patients at the primary-care

http://dx.doi.org/10.5772/63396

level, especially in determining whether a patient should be referred to an appropriate treatment center or not.

11.7. Prevention

The following measures are to be taken so as to prevent and control complications of malaria-related anemia:

1. Follow-up for the patients to decide on the response to treatment.
2. Control of complications of malaria-related anemia.
3. Close monitoring for the possible complications of malaria related anemia especially for cardiac and respiratory complications.
4. Monitoring for selected treatment methods such as adverse drug reactions in antimalarial therapy.
5. Intermittent prophylactic treatment for pregnant women as per the WHO recommendations [23].

12. Anemia in schistosomiasis

Schistosomiasis is considered to fall just second to malaria upon discussing the prevalence of parasitic infestations in the world, being prevalent in more than 70 countries worldwide, with an infection rate affecting one in each 30 individuals. It is most prevalent in tropical and subtropical areas of South America, Africa, and Asia. World Health Organization (WHO) estimates the disease burden to be more than 240 million people infected worldwide, with 400–600 millions of people at risk [24, 25]. Schistosomiasis tends to involve a number of organs leading to dysfunction of these particular organs, such as renal and bladder dysfunction (*Schistosoma haematobium*) or liver and intestinal disease caused by *Schistosoma* (*mansoni, japonicum, mekongi,* and *intercalatum*) in endemic areas, and it is also a contributory cause of anemia and stunting of growth. Schistosomiasis is acquired when cercaria breaks through the skin while swimming or bathing in fresh water when the human host comes into contact with the infectious, free-living, cercarial larvae that are released by the parasite's intermediate hosts [25]. Therefore, patterns of water supply, sanitation, and human water use are critical factors in defining the risk of infection [25, 26]. Moreover, the geographic distribution of the different *Schistosoma* species depends solely on the distribution of the particular snail species that serve as intermediate hosts. On the other hand, the distribution of snails depends on climate, water quality, and other ecologic factors [27, 28]. Thanks to animal models of schistosome infection, where they have allowed intensive study of the immunology and molecular biology of schistosomiasis. Analysis of host responses to these complex multicellular parasites has granted considerable awareness about the regulation of cell-mediated and humoral immunity [29], as well as the resistance pathways available for elimination of macroparasites [30]. Molecular studies of the parasite have granted information on new manners of genetic expression, not only but even the whole genome of the parasite has been sequenced [31, 32], as well as leads for the development of vaccines and new pharmaceuticals for control of this prevalent chronic infection [25].

12.1. Schistosomiasis-related anemia: molecular and genetic basis

The most common presentation of chronic infestation with *S. mansoni* is with the relatively asymptomatic intestinal form of the disease, while a minority develops hepatosplenomegaly characterized by severe hepatic disease complicated by portal hypertension. Such distinct heterogeneity of disease severity is seen among both, humans and experimental mouse model. Severe disease is featured by profound hepatic egg-evoked granulomatous inflammation in a proinflammatory cytokine setting, whereas mild disease conforms with reduced liver inflammation in a Th2 distorted cytokine setting. This distinct variation reflects that genetic differences play a pivotal role in disease development. Smith et al. demonstrated in their study, that severe hepatic pathology in F2 mice 7 wk after infection was significantly associated with a surge in the synthesis of the proinflammatory cytokines IL-17, IFN-γ, and TNF-α by schistosome egg antigen-evoked mesenteric lymph node cells. A quantitative analysis of trait loci revealed a number of genetic intervals governing immunopathology along with IL-17 and IFN-γ production. Egg granuloma size was found to have a significant linkage to the dominantly inherited loci; D4Mit203 and D17Mit82. Moreover, these genetic loci were found to have a decisive effect on the development of immunopathology in murine as evidenced by the significantly reduced hepatic granulomatous inflammation and IL-17 synthesis in interval-specific congenic mice [33]. It is likely due to these genetic differences that a minority of infected persons tends to show the severe form of schistosomiasis with hepatosplenic involvement and hypersplenism.

12.2. Epidemiology of schistosomiasis-associated anemia

In addition to hookworm anemia, anemia in schistosomiasis poses an important public health problem, particularly for those tropical countries in Africa where schistosomiasis is endemic and a strong correlation is found between it and anemia.

12.3. Pathophysiology and manifestations of schistosomiasis-associated anemia

Schistosomiasis or bilharziasis is a group of helminthic infestations that are brought about by blood flatworms of the Schistosoma genus. The pathology of schistosomiasis is typically evoked by ova trapped in the tissues, where the activation of CD4 T cell-mediated immunity results in granulomatous inflammation. Three important forms of schistosomiasis have been described: intestinal, urinary, and hepatic. The former two forms of schistosomiasis are the two common forms relating to anemia. It has been noted that there are several negative effects of the mentioned two forms of schistosomiasis on the coming nutritional parameters in humans [25]:

1. Urinary and fecal blood and iron loss
2. Anemia and hemoglobin levels
3. Proteinuria
4. Child growth and adult protein-energy status
5. Physical fitness and physical activity

It is well known that schistosomiasis can cause iron deficiency anemia by direct blood loss in case of urinary and gastrointestinal schistosomiasis through urine and stools [34]. Interestingly, the hemoglobin level and the hematocrit were found to be inversely related to egg count, in contrary to the prevalence of anemia which tends to increase with increasing egg count [35]. There for it is concluded that this negative association between the degree of infection by *S. haematobium* and iron status showed a deleterious consequence of urinary schistosomiasis on nutrition and hematopoietic status, a thing that should be put in consideration when designing nutrition intervention programs [35]. Other explanations for the anemia associated with schistosomiasis are the anemia of inflammation and the presence of coinfection with other parasites such as hook worm [36, 37].

12.4. Schistosomiasis-related anemia: diagnosis and management

The diagnosis of anemia in schistosomiasis needs evidence of coexistence of both anemia validated by the measurement of hemoglobin and blood fluke infestation by stool or urinary examination for detection of blood fluke ova. Nevertheless, great care should be taken because not all cases with both anemia and blood fluke infestation can be attributed to blood fluke infestation as anemia can be a common copresentation with helminthic infestation in tropical countries. Other etiologies for iron deficiency anemia, particularly hookworm infestation, should be evaluated. Undoubtedly, the coinfestation between hookworm and blood fluke is reported to coexist in the tropics. Compared to hookworm anemia, treatment of anemia in schistosomiasis is usually started with an antihelminthic drug. It has been found that a blanket coverage of a single-dose anthelminthic treatment covering the at-risk population like school children in the endemic areas achieved hematological benefits among most of the children with *S. haematobium* infestation [38]. Such a recommendation is yet waiting establishment in the case of pregnant women [39]. The drug of choice for treatment of schistosomiasis infection is praziquantel (40 mg/kg), similar to other fluke infestations. Moreover, the nutritional supplementation therapy should be similar to hookworm anemia. Nevertheless, as praziquantel does not reduce the hookworm intensity of infection, which is another major cause of anemia in the endemic area, alterations in the prevalence of anemia among the population should be due only to the elimination of *Schistosoma* species infestation. Accordingly, in the area with high prevalence of mixed infection of hookworms and blood flukes, combined antihelminthic drugs for both infestations are advised. It has been found by Friis et al. that the combination of multi-micronutrient fortification and anthelminthic treatment independently raised the hemoglobin. The treatment effect was thought to be due to decrement in *S. mansoni* and hookworm load of infection [40]. However, meta-analysis on this issue did not support their findings, but rather suggested further research on the subject [41]. It has been noted that in areas with schistosomiasis and hookworm infestations, combination treatment with praziquantel and albendazole, plus iron supplementation, is likely to promote good population health and improve hemoglobin levels.

12.5. Schistosomiasis-related anemia: prevention

It is advisable to implement community-level treatment and control of schistosomiasis in endemic areas where protein-energy malnutrition and anemia frequently coexist where such strategies will likely improve child growth, appetite, physical fitness, and activity levels and decrease anemia and symptoms of the infestation [42]. Therefore, screening and early management of identified cases are the best means to prevent schistosomiasis-associated anemia. The development of vaccines will give the solution to this dilemma [43].

Author details

Gasim I Gasim[1] and Ishag Adam[2*]

*Address all correspondence to: ishagadam@hotmail.com

1 Alneelain School of Medicine, Alneelain University, Khartoum, Sudan

2 Faculty of Medicine, University of Khartoum, Khartoum, Sudan

References

[1] Chareonviriyaphap T, Bangs MJ, Ratanatham S. Status of malaria in Thailand. Southeast Asian J Trop Med Public Health. 2000;31:225–37.

[2] Thimasarn K, Jatapadma S, Vijaykadga S, Sirichaisinthop J, Wongsrichanalai C. Epidemiology of Malaria in Thailand. J Travel Med. 1995;2:59–65.

[3] Ouma C, Davenport GC, Garcia S, Kempaiah P, Chaudhary A, Were T, Anyona SB, Raballah E, Konah SN, Hittner JB, Vulule JM, Ong'echa JM, Perkins DJ. Functional haplotypes of Fc gamma (Fcγ) receptor (FcγRIIA and FcγRIIIB) predict risk to repeated episodes of severe malarial anemia and mortality in Kenyan children. Hum Genet. 2012;131(2):289–99.

[4] McGuire W, Knight JC, Hill AV, Allsopp CE, Greenwood BM, Kwiatkowski D. Severe malarial anemia and cerebral malaria are associated with different tumor necrosis factor promoter alleles. J Infect Dis. 1999;179:287–90.

[5] Okeyo WA, Munde EO, Okumu W, Raballah E, Anyona SB, Vulule JM, Ong'echa JM, Perkins DJ, Ouma C. Interleukin (IL)-13 promoter polymorphisms (-7402 T/G and -4729G/A) condition susceptibility to pediatric severe malarial anemia but not circulating IL-13 levels. BMC Immunol. 2013;14:15.

[6] Roberts DJ. Anemia in malaria. In: Post TW, Schrier SL, Daily J, Tirnauer JS, Baron EL (Eds.), *UptoDate*.27th of January 2015. Available from: http://www.uptodate.com/contents/anemia-in-malaria

[7] Sexton AC, Good RT, Hansen DS, D'Ombrain MC, Buckingham L, Simpson K, Schofield L. Transcriptional profiling reveals suppressed erythropoiesis, up-regulated glycolysis, and interferon-associated responses in murine malaria. J Infect Dis. 2004;189:1245–56.

[8] Keller CC, Yamo O, Ouma C, et al. Acquisition of hemozoin by monocytes down-regulates interleukin- 12 p40 (IL-12p40) transcripts and circulating IL-12p70m through an IL-10-dependent mechanism: in vivo and in vitro findings in severe malaria related anemia. Infect Immun. 2006;74:5249–60.

[9] Gourley IS, Kurtis JD, Kamoun M, Amon JJ, Duffy PE. Profound bias in interferon-gamma and interleukin-6 allele frequencies in western Kenya, where severe malaria related anemia is common in children. J Infect Dis 2002;186:1007–12.

[10] Wiwanitkit V. Tropical anemia. Nova Science Publishers, Inc.,Nova Science Publishers, Inc.400 Oser Ave Suite 1600Hauppauge NY 11788-3619United States of AmericaPh: (631)231-7269 Fax: (631)231-8175Email: HYPERLINK "mailto:nova.main@novapublishers.com" nova.main@novapublishers.com

[11] Castro-Gomes T, Mourão LC, Melo GC, Monteiro WM, Lacerda MV, Braga ÉM. Potential immune mechanisms associated with anemia in Plasmodium vivax malaria: a puzzling question. Infect Immun. 2014;82(10):3990–4000.

[12] Olutola A, Mokuolu O. . Severe malaria anaemia in children. In: Silverberg D (Ed.), Anemia. ISBN: 978-953-51-0138-3, InTech. 27th of January 2012. Available from: http://www.intechopen.com/books/anemia/severe-malaria-anaemia-in-children InTech Europe University Campus STeP Ri Slavka Krautzeka 83/A 51000 Rijeka, Croatia Phone: +385 (51) 770 447 Fax: +385 (51) 686 166 HYPERLINK "http://www.intechopen.com" www.intechopen.com InTech China Unit 405, Office Block, Hotel Equatorial Shanghai No.65, Yan An Road (West), Shanghai, 200040, China Phone: +86-21-62489820 Fax: +86-21-62489821

[13] Chan JA, Fowkes FJ, Beeson JG. Surface antigens of Plasmodium falciparum-infected erythrocytes as immune targets and malaria vaccine candidates. Cell Mol Life Sci. 2014;71(19):3633–57.

[14] Balaji SN, Trivedi V. Extracellular methemoglobin mediated early ROS spike triggers osmotic fragility and RBC destruction: an insight into the enhanced hemolysis during malaria. Indian J Clin Biochem. 2012;27(2):178–85.

[15] Perkins DJ, Were T, Davenport GC, Kempaiah P, Hittner JB, Ong'echa JM. Severe malarial anemia: innate immunity and pathogenesis. Int J Biol Sci. 2011;7(9):1427–42.

[16] Ihekwereme CP, Esimone CO, Nwanegbo EC. Hemozoin inhibition and control of clinical malaria. Adv Pharmacol Sci. 2014;2014:984150.

[17] Menon MP, Yoon SS. Uganda malaria indicator survey technical working group. Prevalence and factors associated with anemia among children under 5 years of age – Uganda, 2009. Am J Trop Med Hyg. 2015;93(3):521–6.

[18] Okebe JU, Yahav D, Shbita R, Paul M. Oral iron supplements for children in malaria-endemic areas. Cochrane Database Syst Rev. 2011;(10):CD006589.

[19] CDC (Centers for Disease Control and Prevention) Blood safety basics. 2013. Available at: http://www.cdc.gov/bloodsafety/basics.html (accessed on 23 January 2016)

[20] Meremikwu MM, Smith HJ. Blood transfusion for treating malarial anaemia. Cochrane Database Syst Rev. 1999;4:Art. No.: CD001475. 2000;(2):CD001475. doi: 10.1002/14651858.CD001475

[21] Naing C, Whittaker MA, Nyunt-Wai V, Reid SA, Wong SF, Mak JW, Tanner M. Malaria and soil-transmitted intestinal helminth co-infection and its effect on anemia: a meta-analysis. Trans R Soc Trop Med Hyg. 2013;107:672–83.

[22] Semenya AA, Sullivan JS, Barnwell JW, Secor WE. Schistosoma mansoni infection impairs antimalaria treatment and immune responses of rhesus macaques infected with mosquito-borne *Plasmodium coatneyi*. Infect Immun. 2012;80(11):3821–7.

[23] WHO. Technical Expert Group meeting on intermittent preventive treatment in pregnancy (IPTp). In: WH. Organization (Ed.), World Health Organization. Geneva, Switzerland, 2007.

[24] CDC (2013) (Centers for Disease Control and Prevention) The burden of schistosomiasis. http://www.cdc.gov/globalhealth/ntd/diseases/schisto_burden.html (accessed on 25 January 2016).

[25] WHO (2014) Status of vaccine research and development of vaccines for schistosomiasis prepared for WHO PD VAC. http://www.who.int/immunization/research/meetings_workshops/Schistosomiasis_VaccineRD_Sept2014.pdf (accessed 26 January 2016)

[26] Steinmann P, Keiser J, Bos R, Tanner M, Utzinger J. Schistosomiasis and water resources development: systematic review, meta-analysis, and estimates of people at risk. Lancet Infect Dis. 2006;6(7):411–25.

[27] Rowel C, Fred B, Betson M, Sousa-Figueiredo JC, Kabatereine NB, Stothard JR. Environmental epidemiology of intestinal schistosomiasis in Uganda: population dynamics of biomphalaria (Gastropoda: Planorbidae) in Lake Albert and Lake Victoria with observations on natural infections with digenetic trematodes. BioMed Res Int. 2015;2015(Article ID 717261):11. doi:10.1155/2015/717261

[28] McCreesh N, Booth M. The effect of increasing water temperatures on Schistosoma mansoni transmission and *Biomphalaria pfeifferi* population dynamics: an agent-based modelling study. PLoS One. 2014;9(7):e101462.

[29] Colley DG, Secor WE. Immunology of human schistosomiasis. Parasite Immunol. 2014;36(8):347–57.

[30] Butterworth AE, Curry AJ, Dunne DW, et al. Immunity and morbidity in human schistosomiasis mansoni. Trop Geogr Med. 1994;46:197.

[31] Parker-Manuel SJ, Ivens AC, Dillon GP, Wilson RA. Gene expression patterns in larval Schistosoma mansoni associated with infection of the mammalian host. PLoS Negl Trop Dis. 2011;5(8):e1274.

[32] Young ND, Jex AR, Li B, Liu S, Yang L, Xiong Z, Li Y, Cantacessi C, Hall RS, Xu X, Chen F, Wu X, Zerlotini A, Oliveira G, Hofmann A, Zhang G, Fang X, Kang Y, Campbell BE, Loukas A, Ranganathan S, Rollinson D, Rinaldi G, Brindley PJ, Yang H, Wang J, Wang J, Gasser RB. Whole-genome sequence of *Schistosoma haematobium*. Nat Genet. 2012;44(2):221–5.

[33] Smith PM, Shainheit MG, Bazzone LE, Rutitzky LI, Poltorak A, Stadecker MJ. Genetic control of severe egg-induced immunopathology and IL-17 production in murine schistosomiasis. J Immunol. 2009;183(5):3317–23.

[34] Laudage G, Schirp J. Schistosomiasis– a rare cause of iron deficiency anemia. Leber Magen Darm. 1996;26:216–8.

[35] Prual A, Daouda H, Develoux M, Sellin B, Galan P, Hercberg S. Consequences of *Schistosoma haematobium* infection on the iron status of schoolchildren in Niger. Am J Trop Med Hyg. 1992;47:291–7.

[36] Butler SE, Muok EM, Montgomery SP, Odhiambo K, Mwinzi PM, Secor WE, Karanja DM. Mechanism of anemia in Schistosoma mansoni-infected school children in Western Kenya. Am J Trop Med Hyg. 2012;87(5):862–7.

[37] Ezeamama AE, McGarvey ST, Acosta LP, Zierler S, Manalo DL, Wu H-W, et al. The synergistic effect of concomitant schistosomiasis, hookworm, and trichuris infections on children's anemia burden. PLoS Negl Trop Dis. 2008;2(6):e245.

[38] Coulibaly JT, N'gbesso YK, Knopp S, Keiser J, N'Goran EK, Utzinger J. Efficacy and safety of praziquantel in preschool-aged children in an area co-endemic for *Schistosoma mansoni* and *S. haematobium*. PLoS Negl Trop Dis. 2012;6(12):e1917.

[39] Salam RA, Haider BA, Humayun Q, Bhutta ZA. Effect of administration of antihelminthics for soil-transmitted helminths during pregnancy. Cochrane Database Syst Rev. 2015;6:CD005547.

[40] Friis H, Mwaniki D, Omondi B, Muniu E, Thiong'o F, Ouma J, Magnussen P, Geissler PW, Fleischer Michaelsen K. Effects on haemoglobin of multi-micronutrient supple-

mentation and multi-helminth chemotherapy: a randomized, controlled trial in Kenyan school children. Eur J Clin Nutr. 2003;57:573–9.

[41] Taylor-Robinson DC, Maayan N, Soares-Weiser K, Donegan S, Garner P. Deworming drugs for soil-transmitted intestinal worms in children: effects on nutritional indicators, haemoglobin, and school performance. Cochrane Database Syst Rev. 2015;7:CD000371.

[42] Belizario VY Jr, Totañes FI, de Leon WU, Lumampao YF, Ciro RN. Soil-transmitted helminth and other intestinal parasitic infections among school children in indigenous people communities in Davao del Norte, Philippines. Acta Trop. 2011;120(Suppl 1):S12–8.

[43] Gasim GI, Bella A, Adam I. Schistosomiasis, hepatitis B and hepatitis C co-infection. Virol J. 2015;12:19.

Section 3

Nutritional Deficiencies and Their Outcomes

Chapter 8

Redox Homeostasis in Neural Plasticity and the Aged Brain

Pablo Muñoz, Francisca García, Carolina Estay, Alejandra Arias, Cecilia Hidalgo and Álvaro O. Ardiles

Additional information is available at the end of the chapter

http://dx.doi.org/10.5772/63931

Abstract

Currently, humans can easily live for 60 years and more. This increase in life expectancy produces myriad changes in our bodies that diminish the individual's physical and mental capacities and affect as well the functional capacity of individuals to interact appropriately with their social and physical environments. The oxidative theory of aging predicts an accumulation of oxidative damage to proteins, lipids, and DNA with age; as a consequence, the aged brain gradually suffers loss in neuronal functions, increasing the risk of developing neurodegenerative diseases and cognitive impairment. To date, there are no effective treatments to prevent age-related cognitive decline, making it urgent to identify the neural mechanisms that are altered during aging. In this chapter, we discuss the mechanisms that underlie synaptic plasticity, emphasizing the relationship between redox balance and neuronal function, and we also address current evidence supporting oxidative stress as an important contributing factor in brain aging.

Keywords: synaptic plasticity, aging, neuronal function, oxidative stress, cognitive decline

1. Introduction

Currently, humans can easily live for 60 years and more. This increase in life expectancy produces myriad changes in our bodies that diminish the individual's physical and mental capacities and that affect as well the functional capacity of the individual to interact appropriately with the environment. To date, there are no effective treatments to deter the age-related cognitive decline.

Thus, in order to identify potential therapeutic targets and thus improve the quality of life of aging individuals it becomes urgent to decipher the neural mechanisms that are altered during aging. This knowledge will allow the design of molecules for the prevention, delay or even the reversal of age-related cognitive malfunction.

During aging, the brain undergoes a progressive accumulation of oxidative damage to macromolecules, such as synaptic proteins, lipids, and DNA, which gradually alters neuronal functions and increases the risk of developing neurodegenerative diseases and cognitive impairments.

As you would expect, various nutritional interventions that somehow prevent or counteract this oxidative accumulation, they have proved effective in mitigating the effects of brain aging. Although there is no consensus, the evidence suggests that the consumption of dietary antioxidants may have a beneficial effect on the mental health of aging individuals by a mechanism that involves improvement of synaptic plasticity (SP) [1], as well as an increase in the flow of blood and neurogenesis [2]. Caloric restriction, a significant decrease in caloric intake, is another nutritional intervention in animal models with positive impact in synaptic plasticity, which produces attenuation in the effects related to aging in the CA3 region of the hippocampus [3, 4].

In particular, aging entails synaptic failure, so to achieve a better understanding of the changes associated with the aging process and how to prevent, it is necessary to understand how it affects the mechanisms underlying synaptic plasticity.

This chapter discusses the following topics:

a. Cellular mechanisms of synaptic plasticity.

b. Redox regulation mechanisms acting in synaptic plasticity.

c. Disturbances of neuronal redox homeostasis and accumulation of oxidative damage during aging.

2. Cellular mechanisms of synaptic plasticity

Synaptic plasticity, defined as functional and structural changes occurring in the synapse in response to specific neuronal activity, is critical for the processing of information by the brain. It is widely accepted that changes in synaptic connections, which represent the cellular basis of memories, are encoded and stored in the central nervous system. At the cellular level, these changes occur when a postsynaptic neuron responds to a given presynaptic stimulation caused by either depolarization or neurotransmitter release. This postsynaptic response initiates a series of metabolic changes that promote, in turn, the stimulation of gene expression necessary to enable and consolidate long-lasting structural and functional changes. In its most general form, the hypothesis linking SP and memory states that neuronal activity-dependent plasticity is induced in certain synapses during memory formation. This fact is necessary and sufficient to store information, outlining the type of memory involved in the area of the brain in which plasticity is observed [5–7].

At the electrophysiological level, SP entails an increase or a decrease in synaptic efficiency; these changes are known as long-term potentiation (LTP) (**Figure 1**) or long-term depression (LTD), respectively [8]. If changes occur on the same synapse that was stimulated, they give rise to homosynaptic plasticity. Alternatively, if changes occur in synapses other than those stimulated, they originate heterosynaptic plasticity [9, 10]. The cellular and molecular mechanisms of LTP and LTD have been extensively studied in the hippocampus [11], an area involved in the formation of spatial memory in rodents and humans. In the hippocampus, many forms of plasticity, including LTP and LTD, require an increase in calcium concentration in the postsynaptic neuron [7, 12, 13].

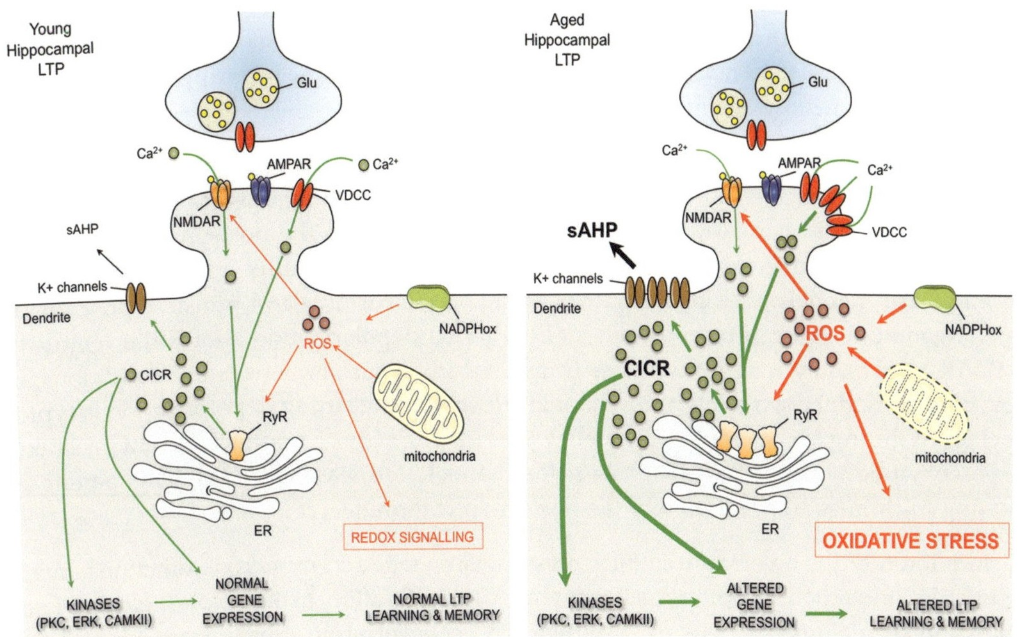

Figure 1. Cartoon depicting the signaling pathways involved in synaptic plasticity. In response to glutamate, NMDA (N-methyl-D-aspartate) receptors (NMDARs) and L-type voltage-dependent calcium channels (VDCCs) open allowing the influx of calcium into the cell. The increase in cytosolic calcium concentration activates the release of calcium from the endoplasmic reticulum (ER) by ryanodine receptors (RyRs) through a calcium-induced calcium-released mechanism (CICR) amplifying calcium signaling. In young synapses, this calcium elevation leads to the activation of calcium-dependent kinases, which are relevant to synaptic plasticity and learning and memory. NMDAR activation also induces the generation of ROS by the activation of NADPH oxidase. Mitochondrial-generated ROS can also contribute to the increase of calcium concentration through the modulation of the activity of NMDARs and RyRs. In aged synapses, the generation of ROS by mitochondria is exacerbated leading to oxidative stress. Ion imbalance, lipoperoxidation, DNA damage, and metabolic impairment are signatures of aged conditions. Increased density of VDCC and calcium-dependent potassium channels produce an increase of the slow component after hyperpolarization (sAHP), contributing to the aging process.

In the hippocampal CA1 area, activity-dependent postsynaptic calcium increments are initiated by calcium influx from the extracellular space through glutamate receptors of the N-methyl-D-aspartate (NMDA) type or through voltage-dependent calcium channels (VDCCs) (**Figure 1**). The resulting calcium signals are amplified and propagated through the calcium-induced calcium release (CICR) mechanism, which engages calcium release from intracellular stores [14]. Much of the experimental work concerning the possible role of LTP in learning has been focused on NMDA receptor (NMDAR)-dependent LTP [15–19]. These studies have shown that pharmacological NMDAR blockage impairs both learning and SP [16, 18–20]. Similarly, spatial memory is impaired in mice with a mutation in the NMDAR R1 subunit; this impairment correlates with alterations in LTP or LTD [21]. Hippocampal NMDAR-dependent SP includes classical associative long-term plasticity between the perforant/dentate gyrus pathway and between neurons in the CA3 and CA1 circuits; the coupling between the excitatory postsynaptic potentials (EPSPs) is associated with action potentials [22]. In addition, post-tetanic potentiation (PTP) and paired pulses facilitation (PPF) are two forms of short-term plasticity, which critically require NMDAR activation [23–25]. A number of NMDAR-independent SP also occur in the hippocampus, including LTP between the mossy fibers of the dentate gyrus and CA3 neurons; this plasticity is induced by the brief application of certain growth factors or neurotransmitters [26, 27].

At glutamatergic synapses, glutamate released from the presynaptic vesicles diffuses across the synaptic cleft and then binds to ionotropic receptors for α-amino-3-hydroxy-5-methylisoxazole-4-propionic acid (AMPA) and to NMDAR present in the postsynaptic membrane. The activation of AMPA receptors (AMPARs) allows sodium influx and potassium efflux through these channels, causing excitatory postsynaptic currents and transient postsynaptic depolarization, which can be recorded as EPSP. This depolarization allows the removal of NMDAR blockage by magnesium ions (Mg^{2+}), which occurs under resting conditions and prevents ion conduction through the channel. However, the strong depolarization is typically achieved during a train of high-frequency presynaptic activity or direct current injection to the postsynaptic neuron, release Mg^{2+} from the channel pore, leading to NMDAR opening and allowing calcium/sodium influx into the postsynaptic terminal [28].

Calcium ion (Ca^{2+}) is a versatile cellular messenger engaged in numerous functions, including muscle contraction, apoptosis, cell growth, cell proliferation, synaptic plasticity, and gene expression regulation [29]. The extracellular Ca^{2+} concentration ranges from 1.5 to 2 mM, whereas the intracellular resting concentration is four orders of magnitude lower and ranges from 70 to 100 nM [30–32]. Increments in neuronal intracellular calcium concentration are of great physiological importance because they play an important role in neuronal functions such as neurotransmitter release, excitability, neuronal plasticity, and gene expression, all functions that are associated with SP and memory formation [30, 31, 33, 34]. In particular, postsynaptic calcium influx triggers amplification and signal propagation by CICR; this process involves the activation of intracellular calcium channels present in endoplasmic reticulum (ER) [14, 30, 34, 35], a continuous membranous network distributed throughout the neuronal cell, including dendrites and dendritic spines, the soma, and the surrounding the nucleus and extending into the axon to reach the presynaptic terminals, where it is closely associated with mitochondria [36] (**Figure 1**).

Two different types of intracellular calcium channels are involved in calcium release from the ER: the inositol 1, 4, 5-trisphosphate (IP3) receptor (IP3R) and the ryanodine receptor (RyR) channels [14, 30, 34, 35]. In areas such as the hippocampus, calcium release is mediated primarily through RyR [37] (**Figure 1**), providing the largest share of the increase of calcium in the postsynaptic spines. The RyR channel has been cloned, purified, and sequenced from several species. Three isoforms of this receptor have been identified: RyR1, RyR2, and RyR3, each one encoded by a different gene. All three RyR isoforms have about 5000 amino acid residues and present 65% of identity among them [38, 39]. RyR1 is the isoform primarily expressed in skeletal muscle, RyR2 in cardiac muscle, and RyR3 in diaphragm muscle [34, 40, 41]. All isoforms are expressed in the brain, RyR2 being the most abundant isoform [41–43]. At the anatomical level, RyR channels have been found in the frontal cortex, olfactory bulb, thalamus, amygdala, cingulate cortex, piriform cortex, entorhinal cortex, occipital cortex, and hippocampus [40–42, 44].

The ryanodine receptor is a homotetramer with a molecular mass of >2 MDa in which each subunit has a weight of ~560 kDa [38, 39]. It has a large N-terminal segment located in the cytoplasm that represents about 90% of the protein, while its C-terminal is small and also facing the cytoplasm. The presence of a large N-terminal site allows its scaffold function for various proteins that play central roles in signal transduction pathways mediated by calcium [45]. The RyR channel can be activated by calcium, caffeine, 4-chloro-m-cresol (4-CMC), and ryanodine, the alkaloid from which its name comes. However, ryanodine has a dual effect on the receptor because at low concentrations (<1 μM) it activates the channel, whereas at high concentrations (>10 μM) it abolishes channel activity [37, 46]. RyR channels are also susceptible to inhibition by dantrolene and are modulated by adenosine triphosphate (ATP), H+, Mg^{2+}, kinases, phosphatases, and reactive oxygen species (ROS) [38].

Different authors have shown that RyR-mediated Ca^{2+} signals have a role in SP, learning, and memory. In primary hippocampal cultures, RyR agonists such as ryanodine and caffeine generate calcium signals that trigger neuritic growth [47] as well as dendritic remodelling [43]. At the electrophysiological level, RyR inhibition by inhibitory concentrations of ryanodine suppresses sustained LTP, while ryanodine concentrations that activate RyR promote hippocampal LTP induction [48]. Moreover, RyR2/RyR3 expression in young rats increases after a hippocampal-dependent behavioral task [43, 44].

Using a genetic approach, it has been demonstrated that the absence of the RyR2 or the RyR3 isoforms, but not of RyR1, impairs spatial memory, while ryanodine concentrations that activate RyR channels promote memory consolidation in rats [39], mice [49], and chickens [50]. By contrast, the RyR inhibitor dantrolene alters spatial memory [35].

3. Reactive oxygen species in synaptic plasticity

As mentioned in the previous section, NMDAR activation is required for various forms of hippocampal plasticity, which produce a number of second messengers, including cAMP, nitric oxide, arachidonic acid, and, certainly, calcium. Recently, ROS have been added to the list of molecules generated upon activation of NMDAR [51, 52].

The role of ROS as synaptic transmission modulators is complex and depends on the concentration and duration of the oxidant stimulus [53, 54]. However, after LTP induction, neuronal ROS generation has been detected [54, 55]. This observation not only suggests that ROS may be necessary for LTP induction but also suggests that ROS generation is important for normal neuronal activity, as discussed subsequently.

Pharmacological activation of NMDAR induces the production of superoxide ($O_2^{\cdot-}$) [51], which raises the question of whether ROS generation meets some physiological function. Ascribing to $O_2^{\cdot-}$ only a neurotoxic effect is not correct, because inhibition of $O_2^{\cdot-}$ production results in significant reduction of LTP induction [54, 55]. The NADPH oxidase complex is a recently characterized superoxide-generating source during NMDAR activation in neurons [52, 56]. This oxidase has been extensively characterized in immune cells [57, 58], but compelling evidence also suggests that superoxide anion generation mediated by this enzyme plays an important role in LTP induction. Incubation of brain slices with superoxide dismutase (SOD), an antioxidant enzyme that catalyzes the removal of superoxide, or with permeable and impermeable superoxide scavengers results in LTP attenuation [55, 59, 60]. Consistent with the above reports, hippocampal LTP is also affected in transgenic mice overexpressing Cu/Zn-SOD isoform [59, 61]. Thus, superoxide anion may be produced as a result of, or in conjunction with, other molecules necessary for LTP. Although superoxide has been implicated to SP-related kinases, such as protein kinase C (PKC) [62] and extracellular signal-regulated kinase (ERK) [63], its mechanism of action is not yet defined.

In addition to the modulating effect of superoxide on SP and memory, hydrogen peroxide (H_2O_2) also affects signaling pathways involved in SP [64]. For instance, H_2O_2 regulates the activity of several kinases, such as PKC and the mitogen-activated protein kinase (MAPK) family [65–67], as well as phosphatases such as calcineurin [66]. Accordingly, incubation of hippocampal slices with catalase, an enzyme that degrades H_2O_2 to water, exhibits altered hippocampal LTP, further supporting a role for H_2O_2 in SP [53, 68, 69].

Remarkably, another electrophysiological study reported conflicting results that are difficult to interpret [53], illustrating the complexity of redox signaling. Using hippocampal slices, it was found that a high concentration of H_2O_2 reduces synaptic responses, while exposure to an intermediate concentration has no apparent effect on the expression of pre-established LTP, but prevents induction of a new LTP. Surprisingly, a low concentration of H_2O_2 increased LTP expression compared to control (absence of peroxide). Another important effect of a low concentration of H_2O_2 is the suppression of LTD [53]. It seems that high concentrations of H_2O_2 can give rise to secondary oxidative reactions unrelated to a physiological response; by contrast, low concentrations of H_2O_2 can be part of the normal mechanism for LTP induction.

Within the context of ROS as cellular messengers, both Fenton and Haber-Weiss reactions are catalyzed by iron and involve superoxide anion and hydrogen peroxide. This feature implies that the mentioned reactions may have biological relevance [70]. While there are few studies that make a direct relation between iron-generated ROS and SP [71, 72], there are clinical data showing that iron deficiency during childhood has a strong impact on some cognitive functions related to SP and that this effect persists in the adult individual despite restoration of normal iron levels [73–75].

http://dx.doi.org/10.5772/63931

4. Redox imbalance in the aged brain

One of the most plausible theories of aging, which involves free radicals, was proposed by Harman in the mid-1950s [76]. This theory suggests that free radicals generated during aerobic metabolism produce cumulative damage in various cellular components, resulting in a loss of function characteristic in the physiology of organisms during aging. Subsequently, this hypothesis was refined including mitochondria as the main source of free radical production in physiological processes [77], as the inner membrane of mitochondria consumes nearly 90% of the oxygen generated by cellular respiration. According to this theory, ROS generated as by-products produce oxidative stress that damages the mitochondria itself, which results in dysfunctional mitochondria with the passing of years.

Particularly, the brain is sensitive to oxidative damage due to several factors. Among them, we can mention its mitochondrial high metabolic rate, the presence of high concentrations of polyunsaturated fatty acids, the presence of transition metals such as iron that are involved in the generation of the hydroxyl radical and the consequent lipid peroxidation, and, finally, a lower antioxidant capacity compared with other organs [78, 79].

This theory predicts that by controlling oxidative stress it is possible to delay the effects of aging. However, studies in *Drosophila melanogaster*, *Caenorhabditis elegans*, and mice, in which the expression of antioxidant genes was experimentally increased, did not achieve the expected impact on lifespan [80–82]. Furthermore, a recent study using post-mortem brains showed no significant differences in glutathione levels in aged individuals compared to younger ones, although a number of increments were found in some brain regions as the frontal and occipital cortex, caudate nucleus, and cerebellum [83]. The fact that this theory cannot fully explain the changes associated with aging shows how complex and multifactorial the aging process is.

Evidence supporting this theory in patients or in Parkinson's and Alzheimer's disease animal models suggests a decrease of glutathione and glutathione transferase activity with age in selected areas of the brain and ventricular cerebrospinal fluid [84]. In this regard, it has been described that ROS and oxidative damage may contribute to cerebral aging and also to a higher prevalence of neurodegenerative diseases. Thus, whereas diverse evidences correlate neuronal changes in redox status with progressive aging in the brain [85–90], others show that the cumulative effect of oxidative stress in neurodegenerative diseases may depend on the organism [91]. Multiple lines of evidence suggest that in the brain changes in the redox environment induce cognitive decline with age, by alterations in synaptic function or intracellular calcium regulation [92, 93]. Aging is associated with impairment in the ability to store, retain, and retrieve information, affecting declarative memory in humans, primates, dogs, and rodents [94–98].

Initially, the deficit in cognitive function associated with aging was partly attributed to a decrease in the number of neurons in the hippocampus [31, 36, 99], a critical region for spatial memory [100–102], or by diminished activity in the prefrontal cortex, the brain area involved in working memory, attention, and planning [95, 103]. However, subsequent studies reported that the loss of neurons does not contribute significantly to the cognitive decline associated with aging; rather, a decrease in synaptic connectivity has been linked to cognitive impairment [99, 104]. On the other hand, numerous evidences suggest that cognitive impairment associated with aging may be due to changes in gene expression [40, 102, 105–109], alteration of calcium homeostasis [32, 93, 110, 111], or changes in the redox state of the cell [14, 34, 86]. All these changes will trigger alterations in the LTP and LTD, two models of SP considered the cellular mechanism of learning and memory [112].

Intracellular calcium deregulation affects a number of synaptic components and calcium-dependent processes [31, 113]. The activity of NMDAR depends on neuronal redox state, so that variations of oxidative stress strongly impact both synaptic responses and NMDAR-dependent plasticity [88, 90]. This property is of great interest since the spatial memory impairment occurred in middle-aged animals is also dependent on NMDAR function [114]. Interestingly, manipulating ROS levels by genetic overexpression of antioxidant enzymes has revealed a direct relationship between the cognitive impairment in aged animals and the regulation of NMDAR [89].

Another synaptic process affected by aging is the increase in the slow component of the Ca^{2+} activated K^+- after hyperpolarization (sAHP) in neurons of the hippocampal CA1 region in aged rodents [115–117]. While it has been described that the sAHP can be modified by an increase in L-type voltage-dependent calcium channels [102, 105] or an increase in calcium-dependent potassium channel density [118], enhanced calcium release from intracellular compartments could also make a significant contribution [85, 111, 119]. In fact, among these factors, the latter becomes crucial because oxidative stress increases during aging and RyR channels are highly sensitive to cellular redox state, increasing their activity in response to oxidation [34, 120, 121].

Certainly, further research is needed in the search to understand and combat the deleterious effects of aging; knowledge gained from these studies should help the development of effective treatments for brain pathologies associated with aging.

Acknowledgements

I like to thank Dr. Mario Párraga for his valuable comments and review of the manuscript.

Author details

Pablo Muñoz[1,2*], Francisca García[1], Carolina Estay[1], Alejandra Arias[3], Cecilia Hidalgo[3,4] and Álvaro O. Ardiles[5]

*Address all correspondence to: pablo.munozca@uv.cl

1 Department of Pathology and Physiology, School of Medicine, Universidad de Valparaíso, Valparaíso, Chile

http://dx.doi.org/10.5772/63931

2 Interdisciplinary Center for Health Innovation, Universidad de Valparaíso, Valparaíso, Chile

3 Biomedical Neuroscience Institute, Faculty of Medicine, Universidad de Chile, Santiago, Chile

4 Institute of Biomedical Sciences and Center for Molecular Studies of the Cell, Faculty of Medicine, Universidad de Chile, Santiago, Chile

5 Interdisciplinary Center for Neuroscience, Universidad de Valparaíso, Valparaíso, Chile

References

[1] Dias, G.P., et al., *The role of dietary polyphenols on adult hippocampal neurogenesis: molecular mechanisms and behavioural effects on depression and anxiety*. Oxid Med Cell Longev, 2012. 2012: p. 541971.

[2] Spencer, J.P., *The impact of fruit flavonoids on memory and cognition*. Br J Nutr, 2010. 104 Suppl 3: pp. S40–7.

[3] Adams, M.M., et al., *Caloric restriction and age affect synaptic proteins in hippocampal CA3 and spatial learning ability*. Exp Neurol, 2008. 211(1): pp. 141–9.

[4] Mladenovic Djordjevic, A., et al., *Long-term dietary restriction modulates the level of presynaptic proteins in the cortex and hippocampus of the aging rat*. Neurochem Int, 2010. 56(2): pp. 250–5.

[5] Kandel, E.R. and J.H. Schwartz, *Molecular biology of learning: modulation of transmitter release*. Science, 1982. 218(4571): pp. 433–43.

[6] Lynch, G. and M. Baudry, *The biochemistry of memory: a new and specific hypothesis*. Science, 1984. 224(4653): pp. 1057–63.

[7] Bailey, C.H., E.R. Kandel, and K.M. Harris, *Structural components of synaptic plasticity and memory consolidation*. Cold Spring Harb Perspect Biol, 2015. 7(7): p. a021758.

[8] Sweatt, J.D., *Neural plasticity & behavior – sixty years of conceptual advances*. J Neurochem, 2016.

[9] Nguyen, P.V., *Heterosynaptic strengthening of hippocampal LTP*. Trends Neurosci, 2001. 24(9): pp. 502–3.

[10] Kirkwood, A. and M.F. Bear, *Homosynaptic long-term depression in the visual cortex*. J Neurosci, 1994. 14(5 Pt 2): pp. 3404–12.

[11] Stanton, P.K., *LTD, LTP, and the sliding threshold for long-term synaptic plasticity*. Hippocampus, 1996. 6(1): pp. 35–42.

[12] Ni, Z., et al., *Heterosynaptic modulation of motor cortical plasticity in human*. J Neurosci, 2014. 34(21): pp. 7314–21, doi: 10.1111/jnc.13580.

[13] Connor, S.A. and Y.T. Wang, *A place at the table: LTD as a mediator of memory genesis*. Neuroscientist, 2015.

[14] Paula-Lima, A.C., T. Adasme, and C. Hidalgo, *Contribution of Ca2+ release channels to hippocampal synaptic plasticity and spatial memory: potential redox modulation*. Antioxid Redox Signal, 2014. 21(6): pp. 892–914.

[15] Baker, K.B. and J.J. Kim, *Effects of stress and hippocampal NMDA receptor antagonism on recognition memory in rats*. Learn Mem, 2002. 9(2): pp. 58–65.

[16] Gould, T.J. and M.C. Lewis, *Coantagonism of glutamate receptors and nicotinic acetylcholinergic receptors disrupts fear conditioning and latent inhibition of fear conditioning*. Learn Mem, 2005. 12(4): pp. 389–98.

[17] Jia, Z., et al., *Selective abolition of the NMDA component of long-term potentiation in mice lacking mGluR5*. Learn Mem, 1998. 5(4–5): pp. 331–43 doi:10.1177/1073858415588498.

[18] Torras-Garcia, M., et al., *Reconsolidation after remembering an odor-reward association requires NMDA receptors*. Learn Mem, 2005. 12(1): pp. 18–22.

[19] Fanselow, M.S., et al., *Differential effects of the N-methyl-D-aspartate antagonist DL-2-amino-5-phosphonovalerate on acquisition of fear of auditory and contextual cues*. Behav Neurosci, 1994. 108(2): pp. 235–40.

[20] Mao, Y., et al., *Early chronic blockade of NR2B subunits and transient activation of NMDA receptors modulate LTP in mouse auditory cortex*. Brain Res, 2006. 1073–1074: pp. 131–8.

[21] Collingridge, G.L. and T.V. Bliss, *Memories of NMDA receptors and LTP*. Trends Neurosci, 1995. 18(2): pp. 54–6.

[22] Park, P., et al., *NMDA receptor-dependent long-term potentiation comprises a family of temporally overlapping forms of synaptic plasticity that are induced by different patterns of stimulation*. Philos Trans R Soc Lond B Biol Sci, 2014. 369(1633): p. 20130131.

[23] Lauri, S.E., et al., *Presynaptic mechanisms involved in the expression of STP and LTP at CA1 synapses in the hippocampus*. Neuropharmacology, 2007. 52(1): pp. 1–11.

[24] Lauri, S.E., et al., *Functional maturation of CA1 synapses involves activity-dependent loss of tonic kainate receptor-mediated inhibition of glutamate release*. Neuron, 2006. 50(3): pp. 415–29.

[25] Klein, T., et al., *The role of heterosynaptic facilitation in long-term potentiation (LTP) of human pain sensation*. Pain, 2008. 139(3): pp. 507–19.

[26] Kirkwood, A., et al., *Modulation of long-term synaptic depression in visual cortex by acetylcholine and norepinephrine*. J Neurosci, 1999. 19(5): pp. 1599–609.

[27] Wibrand, K., et al., *Identification of genes co-upregulated with Arc during BDNF-induced long-term potentiation in adult rat dentate gyrus in vivo*. Eur J Neurosci, 2006. 23(6): pp. 1501–11.

[28] Emptage, N., T.V. Bliss, and A. Fine, *Single synaptic events evoke NMDA receptor-mediated release of calcium from internal stores in hippocampal dendritic spines*. Neuron, 1999. 22(1): pp. 115–24.

[29] Berridge, M.J., P. Lipp, and M.D. Bootman, *Signal transduction. The calcium entry pas de deux*. Science, 2000. 287(5458): pp. 1604–5.

[30] Berridge, M.J., P. Lipp, and M.D. Bootman, *The versatility and universality of calcium signalling*. Nat Rev Mol Cell Biol, 2000. 1(1): pp. 11–21.

[31] Foster, T.C., *Calcium homeostasis and modulation of synaptic plasticity in the aged brain*. Aging Cell, 2007. 6(3): pp. 319–25.

[32] Hidalgo, C. and M.A. Carrasco, *Redox control of brain calcium in health and disease*. Antioxid Redox Signal, 2011. 14(7): pp. 1203–7.

[33] Silva, A.J., et al., *CREB and memory*. Annu Rev Neurosci, 1998. 21: pp. 127–48.

[34] Hidalgo, C., et al., *Redox regulation of RyR-mediated Ca2+ release in muscle and neurons*. Biol Res, 2004. 37(4): pp. 539–52.

[35] Edwards, T.M. and N.S. Rickard, *Pharmaco-behavioural evidence indicating a complex role for ryanodine receptor calcium release channels in memory processing for a passive avoidance task*. Neurobiol Learn Mem, 2006. 86(1): pp. 1–8.

[36] Berridge, M.J., *The endoplasmic reticulum: a multifunctional signaling organelle*. Cell Calcium, 2002. 32(5–6): pp. 235–49.

[37] Fill, M. and J.A. Copello, *Ryanodine receptor calcium release channels*. Physiol Rev, 2002. 82(4): pp. 893–922.

[38] Lanner, J.T., et al., *Ryanodine receptors: structure, expression, molecular details, and function in calcium release*. Cold Spring Harb Perspect Biol, 2010. 2(11): p. a003996.

[39] Van Petegem, F., *Ryanodine receptors: structure and function*. J Biol Chem, 2012. 287(38): pp. 31624–32.

[40] Furuichi, T., et al., *Multiple types of ryanodine receptor/Ca2+ release channels are differentially expressed in rabbit brain*. J Neurosci, 1994. 14(8): pp. 4794–805.

[41] Hertle, D.N. and M.F. Yeckel, *Distribution of inositol-1,4,5-trisphosphate receptor isotypes and ryanodine receptor isotypes during maturation of the rat hippocampus*. Neuroscience, 2007. 150(3): pp. 625–38.

[42] Sharp, A.H., et al., *Differential immunohistochemical localization of inositol 1,4,5-trisphosphate- and ryanodine-sensitive Ca2+ release channels in rat brain*. J Neurosci, 1993. 13(7): pp. 3051–63.

[43] Adasme, T., et al., *Involvement of ryanodine receptors in neurotrophin-induced hippocampal synaptic plasticity and spatial memory formation*. Proc Natl Acad Sci U S A, 2011. 108(7): pp. 3029–34.

[44] Zhao, W., et al., *Spatial learning induced changes in expression of the ryanodine type II receptor in the rat hippocampus*. FASEB J, 2000. 14(2): pp. 290–300.

[45] Berridge, M.J., M.D. Bootman, and H.L. Roderick, *Calcium signalling: dynamics, homeostasis and remodelling*. Nat Rev Mol Cell Biol, 2003. 4(7): pp. 517–29.

[46] Zucchi, R. and S. Ronca-Testoni, *The sarcoplasmic reticulum Ca2+ channel/ryanodine receptor: modulation by endogenous effectors, drugs and disease states*. Pharmacol Rev, 1997. 49(1): pp. 1–51.

[47] Korkotian, E. and M. Segal, *Release of calcium from stores alters the morphology of dendritic spines in cultured hippocampal neurons*. Proc Natl Acad Sci U S A, 1999. 96(21): pp. 12068–72.

[48] Lu, Y.F. and R.D. Hawkins, *Ryanodine receptors contribute to cGMP-induced late-phase LTP and CREB phosphorylation in the hippocampus*. J Neurophysiol, 2002. 88(3): pp. 1270–8.

[49] Galeotti, N., et al., *Different involvement of type 1, 2, and 3 ryanodine receptors in memory processes*. Learn Mem, 2008. 15(5): pp. 315–23.

[50] Baker, K.D., T.M. Edwards, and N.S. Rickard, *A ryanodine receptor agonist promotes the consolidation of long-term memory in young chicks*. Behav Brain Res, 2010. 206(1): pp. 143–6.

[51] Bindokas, V.P., et al., *Superoxide production in rat hippocampal neurons: selective imaging with hydroethidine*. J Neurosci, 1996. 16(4): pp. 1324–36.

[52] Kishida, K.T., et al., *NADPH oxidase is required for NMDA receptor-dependent activation of ERK in hippocampal area CA1*. J Neurochem, 2005. 94(2): pp. 299–306.

[53] Kamsler, A. and M. Segal, *Hydrogen peroxide modulation of synaptic plasticity*. J Neurosci, 2003. 23(1): pp. 269–76.

[54] Knapp, L.T. and E. Klann, *Role of reactive oxygen species in hippocampal long-term potentiation: contributory or inhibitory?* J Neurosci Res, 2002. 70(1): pp. 1–7.

[55] Thiels, E., et al., *Impairment of long-term potentiation and associative memory in mice that overexpress extracellular superoxide dismutase*. J Neurosci, 2000. 20(20): pp. 7631–9.

[56] Tejada-Simon, M.V., et al., *Synaptic localization of a functional NADPH oxidase in the mouse hippocampus*. Mol Cell Neurosci, 2005. 29(1): pp. 97–106.

[57] Babior, B.M., *NADPH oxidase*. Curr Opin Immunol, 2004. 16(1): pp. 42–7.

[58] Cross, A.R. and A.W. Segal, *The NADPH oxidase of professional phagocytes – prototype of the NOX electron transport chain systems*. Biochim Biophys Acta, 2004. 1657(1): pp. 1–22.

[59] Hu, D., E. Klann, and E. Thiels, *Superoxide dismutase and hippocampal function: age and isozyme matter*. Antioxid Redox Signal, 2007. 9(2): pp. 201–10.

[60] Klann, E., *Cell-permeable scavengers of superoxide prevent long-term potentiation in hippocampal area CA1*. J Neurophysiol, 1998. 80(1): pp. 452–7.

[61] Serrano, F. and E. Klann, *Reactive oxygen species and synaptic plasticity in the aging hippocampus*. Ageing Res Rev, 2004. 3(4): pp. 431–43.

[62] Knapp, L.T. and E. Klann, *Potentiation of hippocampal synaptic transmission by superoxide requires the oxidative activation of protein kinase C*. J Neurosci, 2002. 22(3): pp. 674–83.

[63] Kanterewicz, B.I., L.T. Knapp, and E. Klann, *Stimulation of p42 and p44 mitogen-activated protein kinases by reactive oxygen species and nitric oxide in hippocampus*. J Neurochem, 1998. 70(3): pp. 1009–16.

[64] Lee, K. and W.J. Esselman, *Inhibition of PTPs by H(2)O(2) regulates the activation of distinct MAPK pathways*. Free Radic Biol Med, 2002. 33(8): pp. 1121–32.

[65] Palumbo, E.J., et al., *Oxidation-induced persistent activation of protein kinase C in hippocampal homogenates*. Biochem Biophys Res Commun, 1992. 187(3): pp. 1439–45.

[66] Klann, E. and E. Thiels, *Modulation of protein kinases and protein phosphatases by reactive oxygen species: implications for hippocampal synaptic plasticity*. Prog Neuropsychopharmacol Biol Psychiatry, 1999. 23(3): pp. 359–76.

[67] Kemmerling, U., et al., *Calcium release by ryanodine receptors mediates hydrogen peroxide-induced activation of ERK and CREB phosphorylation in N2a cells and hippocampal neurons*. Cell Calcium, 2007. 41(5): pp. 491–502.

[68] Kamsler, A. and M. Segal, *Paradoxical actions of hydrogen peroxide on long-term potentiation in transgenic superoxide dismutase-1 mice*. J Neurosci, 2003. 23(32): pp. 10359–67.

[69] Yermolaieva, O., et al., *Reactive oxygen species and nitric oxide mediate plasticity of neuronal calcium signaling*. Proc Natl Acad Sci U S A, 2000. 97(1): pp. 448–53.

[70] Thannickal, V.J. and B.L. Fanburg, *Reactive oxygen species in cell signaling*. Am J Physiol Lung Cell Mol Physiol, 2000. 279(6): pp. L1005-28.

[71] Munoz, P., et al., *Iron mediates N-methyl-D-aspartate receptor-dependent stimulation of calcium-induced pathways and hippocampal synaptic plasticity*. J Biol Chem, 2011. 286(15): pp. 13382–92.

[72] Munoz, P., et al., *Effect of iron on the activation of the MAPK/ERK pathway in PC12 neuroblastoma cells*. Biol Res, 2006. 39(1): pp. 189–90.

[73] McEchron, M.D. and M.D. Paronish, *Perinatal nutritional iron deficiency reduces hippocampal synaptic transmission but does not impair short- or long-term synaptic plasticity*. Nutr Neurosci, 2005. 8(5–6): pp. 277–85.

[74] Jorgenson, L.A., et al., *Fetal iron deficiency disrupts the maturation of synaptic function and efficacy in area CA1 of the developing rat hippocampus*. Hippocampus, 2005. 15(8): pp. 1094–102.

[75] Jorgenson, L.A., J.D. Wobken, and M.K. Georgieff, *Perinatal iron deficiency alters apical dendritic growth in hippocampal CA1 pyramidal neurons*. Dev Neurosci, 2003. 25(6): pp. 412–20.

[76] Harman, D., *Aging: a theory based on free radical and radiation chemistry*. J Gerontol, 1956. 11(3): pp. 298–300.

[77] Harman, D., *The biologic clock: the mitochondria?* J Am Geriatr Soc, 1972. 20(4): pp. 145–7.

[78] Droge, W., *Oxidative stress and aging*. Adv Exp Med Biol, 2003. 543: pp. 191–200.

[79] Jovanovic, Z. and S. Jovanovic, *[Resistance of nerve cells to oxidative injury]*. Med Pregl, 2011. 64(7–8): pp. 386–91.

[80] Orr, W.C. and R.S. Sohal, *Extension of life-span by overexpression of superoxide dismutase and catalase in Drosophila melanogaster*. Science, 1994. 263(5150): pp. 1128–30.

[81] Larsen, P.L., *Aging and resistance to oxidative damage in Caenorhabditis elegans*. Proc Natl Acad Sci U S A, 1993. 90(19): pp. 8905–9.

[82] Huang, T.T., et al., *Ubiquitous overexpression of CuZn superoxide dismutase does not extend life span in mice*. J Gerontol A Biol Sci Med Sci, 2000. 55(1): pp. B5–9.

[83] Tong, J., et al., *Do glutathione levels decline in aging human brain?* Free Radic Biol Med, 2016. 93: pp. 110–117.

[84] Mazzetti, A.P., et al., *Glutathione transferases and neurodegenerative diseases*. Neurochem Int, 2015. 82: pp. 10–8.

[85] Bodhinathan, K., A. Kumar, and T.C. Foster, *Redox sensitive calcium stores underlie enhanced after hyperpolarization of aged neurons: role for ryanodine receptor mediated calcium signaling*. J Neurophysiol, 2010. 104(5): pp. 2586–93.

[86] Bodhinathan, K., A. Kumar, and T.C. Foster, *Intracellular redox state alters NMDA receptor response during aging through Ca2+/calmodulin-dependent protein kinase II*. J Neurosci, 2010. 30(5): pp. 1914–24.

[87] Ghosh, D., et al., *A reversible early oxidized redox state that precedes macromolecular ROS damage in aging nontransgenic and 3xTg-AD mouse neurons*. J Neurosci, 2012. 32(17): pp. 5821–32.

[88] Haxaire, C., et al., *Reversal of age-related oxidative stress prevents hippocampal synaptic plasticity deficits by protecting D-serine-dependent NMDA receptor activation*. Aging Cell, 2012. 11(2): pp. 336–44.

[89] Lee, W.H., et al., *Role of antioxidant enzymes in redox regulation of N-methyl-D-aspartate receptor function and memory in middle-aged rats*. Neurobiol Aging, 2014. 35(6): pp. 1459–68.

[90] Robillard, J.M., et al., *Glutathione restores the mechanism of synaptic plasticity in aged mice to that of the adult*. PLoS One, 2011. 6(5): pp. e20676.

[91] Salmon, A.B., A. Richardson, and V.I. Perez, *Update on the oxidative stress theory of aging: does oxidative stress play a role in aging or healthy aging?* Free Radic Biol Med, 2010. 48(5): pp. 642–55.

[92] Belrose, J.C., et al., *Loss of glutathione homeostasis associated with neuronal senescence facilitates TRPM2 channel activation in cultured hippocampal pyramidal neurons*. Mol Brain, 2012. 5: pp. 11.

[93] Foster, T.C., *Dissecting the age-related decline on spatial learning and memory tasks in rodent models: N-methyl-D-aspartate receptors and voltage-dependent Ca2+ channels in senescent synaptic plasticity*. Prog Neurobiol, 2012. 96(3): pp. 283–303.

[94] Albert, M.S., *The ageing brain: normal and abnormal memory*. Philos Trans R Soc Lond B Biol Sci, 1997. 352(1362): pp. 1703–9.

[95] Yankner, B.A., T. Lu, and P. Loerch, *The aging brain*. Annu Rev Pathol, 2008. 3: pp. 41–66.

[96] Burke, S.N., et al., *Pattern separation deficits may contribute to age-associated recognition impairments*. Behav Neurosci, 2010. 124(5): pp. 559–73.

[97] Bergado, J.A., et al., *Spatial and emotional memory in aged rats: a behavioral-statistical analysis*. Neuroscience, 2011. 172: pp. 256–69.

[98] Foster, T.C., R.A. Defazio, and J.L. Bizon, *Characterizing cognitive aging of spatial and contextual memory in animal models*. Front Aging Neurosci, 2012. 4: p. 12.

[99] Burke, S.N. and C.A. Barnes, *Neural plasticity in the ageing brain*. Nat Rev Neurosci, 2006. 7(1): pp. 30–40.

[100] Morris, R., *Developments of a water-maze procedure for studying spatial learning in the rat*. J Neurosci Methods, 1984. 11(1): pp. 47–60.

[101] Rapp, P.R., R.A. Rosenberg, and M. Gallagher, *An evaluation of spatial information processing in aged rats*. Behav Neurosci, 1987. 101(1): pp. 3–12.

[102] Veng, L.M. and M.D. Browning, *Regionally selective alterations in expression of the alpha(1D) subunit (Ca(v)1.3) of L-type calcium channels in the hippocampus of aged rats*. Brain Res Mol Brain Res, 2002. 107(2): pp. 120–7.

[103] Agster, K.L. and R.D. Burwell, *Hippocampal and subicular efferents and afferents of the perirhinal, postrhinal, and entorhinal cortices of the rat*. Behav Brain Res, 2013. 254: pp. 50–64.

[104] Rapp, P.R., et al., *Neuron number in the parahippocampal region is preserved in aged rats with spatial learning deficits*. Cereb Cortex, 2002. 12(11): pp. 1171–9.

[105] Thibault, O. and P.W. Landfield, *Increase in single L-type calcium channels in hippocampal neurons during aging*. Science, 1996. 272(5264): pp. 1017–20.

[106] Monti, B., C. Berteotti, and A. Contestabile, *Dysregulation of memory-related proteins in the hippocampus of aged rats and their relation with cognitive impairment*. Hippocampus, 2005. 15(8): pp. 1041–9.

[107] Moyer, J.R., Jr., et al., *Aging-related changes in calcium-binding proteins in rat perirhinal cortex*. Neurobiol Aging, 2011. 32(9): pp. 1693–706.

[108] Bruno, A.M., et al., *Altered ryanodine receptor expression in mild cognitive impairment and Alzheimer's disease*. Neurobiol Aging, 2012. 33(5): p.p 1001 e1–6.

[109] Segal, M. and E. Korkotian, *Endoplasmic reticulum calcium stores in dendritic spines*. Front Neuroanat, 2014. 8: pp. 64.

[110] Kumar, A. and T.C. Foster, *Intracellular calcium stores contribute to increased susceptibility to LTD induction during aging*. Brain Res, 2005. 1031(1): pp. 125–8.

[111] Gant, J.C., et al., *Early and simultaneous emergence of multiple hippocampal biomarkers of aging is mediated by Ca2+ induced Ca2+ release*. J Neurosci, 2006. 26(13): pp. 3482–90.

[112] Citri, A. and R.C. Malenka, *Synaptic plasticity: multiple forms, functions, and mechanisms*. Neuropsychopharmacology, 2008. 33(1): pp. 18–41.

[113] Kumar, A., K. Bodhinathan, and T.C. Foster, *Susceptibility to calcium dysregulation during brain aging*. Front Aging Neurosci, 2009. 1: p. 2.

[114] Kumar, A. and T.C. Foster, *Linking redox regulation of NMDAR synaptic function to cognitive decline during aging*. J Neurosci, 2013. 33(40): pp. 15710–5.

[115] Thibault, O., J.C. Gant, and P.W. Landfield, *Expansion of the calcium hypothesis of brain aging and Alzheimer's disease: minding the store*. Aging Cell, 2007. 6(3): pp. 307–17.

[116] Tombaugh, G.C., W.B. Rowe, and G.M. Rose, *The slow afterhyperpolarization in hippocampal CA1 neurons covaries with spatial learning ability in aged Fisher 344 rats*. J Neurosci, 2005. 25(10): pp. 2609–16.

[117] Matthews, E.A., J.M. Linardakis, and J.F. Disterhoft, *The fast and slow afterhyperpolarizations are differentially modulated in hippocampal neurons by aging and learning*. J Neurosci, 2009. 29(15): pp. 4750–5.

[118] Power, J.M., et al., *Age-related enhancement of the slow outward calcium-activated potassium current in hippocampal CA1 pyramidal neurons in vitro*. J Neurosci, 2002. 22(16): pp. 7234–43.

[119] Kumar, A. and T.C. Foster, *Enhanced long-term potentiation during aging is masked by processes involving intracellular calcium stores*. J Neurophysiol, 2004. 91(6): pp. 2437–44.

[120] Bull, R., et al., *Ischemia enhances activation by Ca2+ and redox modification of ryanodine receptor channels from rat brain cortex*. J Neurosci, 2008. 28(38): pp. 9463–72.

[121] Eager, K.R. and A.F. Dulhunty, *Activation of the cardiac ryanodine receptor by sulfhydryl oxidation is modified by Mg2+ and ATP*. J Membr Biol, 1998. 163(1): pp. 9–18.

Chapter 9

Vitamin D Deficiency

Naji J. Aljohani

Additional information is available at the end of the chapter

http://dx.doi.org/10.5772/63638

Abstract

Previously, known actions of vitamin D were confined to skeletal health, but accumulating evidence has consistently suggested that vitamin D has pleomorphic roles in overall human physiology. Hence, no other micronutrient deficiency in the modern times has gained as much global attention as vitamin D deficiency. In this chapter, the author reinforces what is already known in vitamin D and highlights several important findings in vitamin D research, with a special focus on one of the most vitamin D-deficient regions in the world, the Middle East, and Saudi Arabia, in particular.

Keywords: vitamin D, deficiency

1. Vitamin D physiology

Vitamin D plays an essential role in the regulation of calcium and phosphorus absorption and metabolism for bone health. Nevertheless, the influence of vitamin D is more than just mineral and skeletal homeostasis. The existence of vitamin D receptors (VDR) in several tissues and organs implies that vitamin D physiology encompasses beyond bone maintenance [1]. Furthermore, the enzyme responsible for the conversion of 25[OH]D to its biologically active form [Vitamin D (1, 25$[OH]_2$D)] has been recognized in several other tissues aside from kidneys with evidence growing that extra renal synthesis of 1, 23$[OH]_2$D may be just as important in regulating the cell growth of cellular differentiation via paracrine or autocrine regulatory mechanisms [2–4]. **Figure 1** shows the schematic overview of vitamin D metabolism that starts in the liver, where vitamin D is hydroxylated to 25(OH)D, the main circulating vitamin D metabolite used for vitamin D deficiency diagnosis [5]. Further hydroxylation of 25(OH)D to 1, 25(OH)D is catalyzed by 1α-hydroxylase which is expressed in multiple tissues and binds to vitamin D receptors that in turn regulates various genes [5].

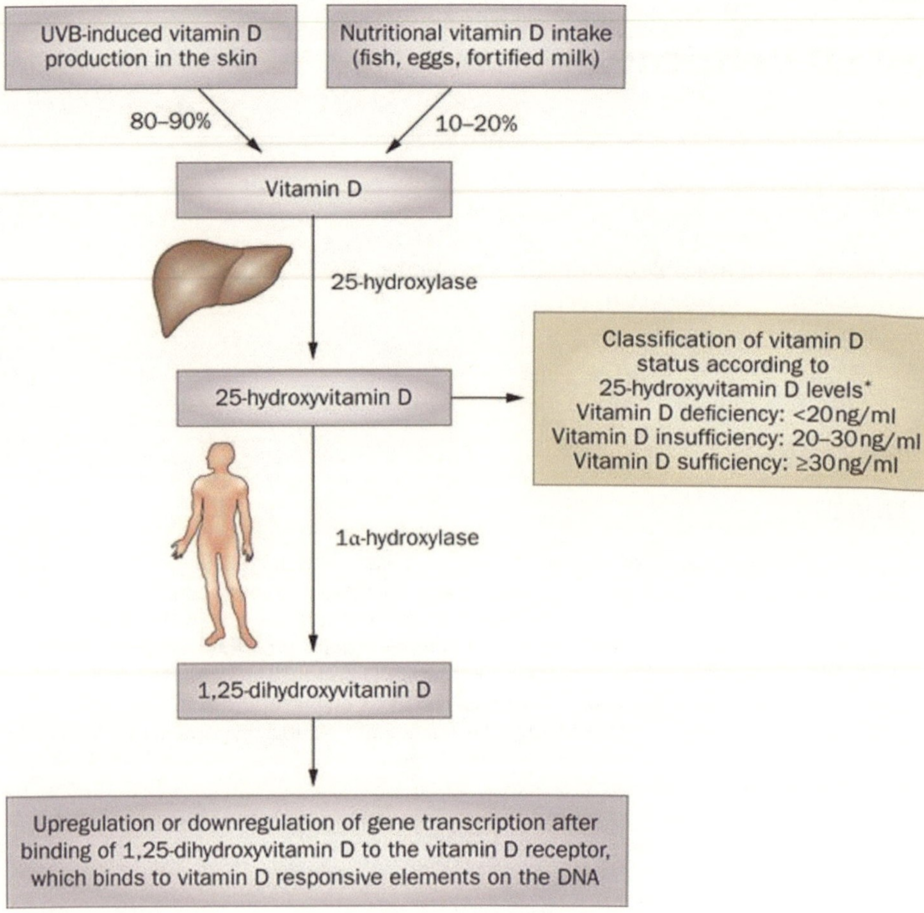

Figure 1. Vitamin D metabolism overview (figure reprinted from Pilz et al. [5]).

Vitamin D's etymology was obtained in the early part of the twentieth century after the discovery of the antirachitic effect of cod liver oil [6]. Then, the unidentified vitamin in cod liver oil was labeled as "D," similar to vitamins A, B, and C, which have been already identified. Synthesis of vitamin D from the skin provides most of the vitamin to the body (80–100%) and with adequate sunlight exposure, dietary vitamin may be unnecessary. However, time spent outdoors or the amount of incidental sun exposure on a regular basis, latitude, age, and skin color influence the cutaneous production of vitamin D and therefore affect vitamin D status. Foods rich in vitamin D include high-fat content fish (sardines, salmon, herring, and mackerel) that are meager and costly, since the study site is situated far from the coast, meat and egg, and fortified milk, juice and margarine. Even in some countries where certain foods are fortified with vitamin D, dietary intake of vitamin D is usually insufficient to maintain adequate levels of 25-hydroxyvitamin D [7]. Currently, there are three treatment modalities for vitamin D deficiency: sunlight, artificial ultraviolet B (UVB) radiation, and vitamin D supplementation [7]. Ideal 25-hydroxyvitamin D [25(OH)D] levels continue to be debated in scientific circles and the definition of vitamin D deficiency changes almost yearly and ranges become higher than previously thought. As of this writing, the most commonly accepted definition of vitamin D deficiency is the one endorsed by the US Endocrine Society, that is, circulating serum 25(OH)D levels <50 nmol/l (<20 ng/ml) [8].

2. Vitamin D deficiency prevalence

Globally, vitamin D deficiency is widespread and is considered as an epidemic [9]. In a systematic literature review done by Hilger et al. in 44 countries involving more than 168,000 participants, 37.3% of the studies reported mean values <50 nmol/l, with the highest values reported in. Furthermore, it was only in the Asia/Pacific and Middle East/African regions where they observed age-related differences [10]. The recent study of Haq et al. also measured vitamin D deficiency prevalence in a single laboratory in United Arab Emirates, and this time involved 60,979 patients coming from 136 countries and revealed severe vitamin deficiency (25(OH)D <25 nmol/l) in 23% of the subjects tested and another 37% falling under mild deficiency (25(OH)D <50 nmol). This study is unique among other large-scale epidemiologic studies since it involved several nationalities in one setting and using only one laboratory, minimizing the need to adjust for known vitamin D cofactors such as geographical location and variability between measurements [11]. The Middle East and North African (MENA) region in general has a very high prevalence of vitamin D deficiency and is most prominent in women of varying ages [12]. The Kingdom of Saudi Arabia (KSA), being part of the MENA region, is not spared from vitamin D deficiency, despite the sunlight-rich environment.

Sedrani [12] was the first to document vitamin D deficiency in KSA, and this was observed among apparently healthy student males of King Saud University, Riyadh, KSA. Since then and in the same year [12], other studies using different healthy subpopulations have emerged, mostly women of child-bearing age [13–16]. In all studies, henceforth, vitamin D deficiency ranged from one out of five Saudis, to almost 100%. Consequently, at this time, rapid industrialization was taking place at KSA. Environmental risk factors in lifestyle such as daytime sleep and night time activities, work environments, which are sedentary and extreme weather conditions, may have been contributory [17]. Certain groups, such as the elderly, dark skinned, and/or veiled women and their children, are at particular risk of hypovitaminosis D [7, 18]. But more importantly, urbanization and tremendous socioeconomic growth has resulted in profound changes in the way of life during the last three decades, resulting in an increased and sustained incidence of obesity and type-2 diabetes mellitus [19], diseases known to elicit depressed circulating levels of vitamin D. As time passed, and with advancing technology and faster dissemination of information, epidemiologic studies on vitamin D deficiency across KSA has emerged. Through the initiatives of HRM, King Abdullah bin Abdulaziz Al-Saud, and the thrust for a knowledge-based economy, the research industry in KSA exponentially flourished and with it, several large scale studies paved way for exposing the worsening vitamin D deficiency in KSA [20, 21]. Furthermore, debilitating diseases associated with vitamin D deficiency have started to emerge and become more prominent, including osteoporosis [22], type-2 diabetes mellitus [23, 24], and systemic lupus erythematosus [25] to name a few.

3. Diseases associated with vitamin D deficiency

Vitamin D deficiency has been consistently associated with hypertension, diabetes mellitus, cardiovascular disease, stroke, multiple sclerosis, inflammatory bowel disease, osteoporosis, periodontal disease, macular degeneration, mental illness, propensity to fall, and chronic pain and various cancers [26]. Most tissues have not only vitamin D receptors, but also hydroxylase enzyme that is required to convert 25(OH)D to the active form, 1α,25-dihydroxyvitamin D3 [27]. Therefore, vitamin D can affect tissues that are not involved in calcium homeostasis and bone metabolism. Almost all tissues in the body possess vitamin D receptors including brain, heart skeletal muscle, smooth muscle cells, pancreas, activated T and B lymphocytes, and monocytes [28].

The major diseases associated with vitamin D deficiency in KSA are listed in **Table 1**. Among these, the most widely documented include vitamin D deficiency rickets among Saudi children and type-1 diabetes mellitus and osteoporosis in adults. It is expected that with the increasing elderly Saudi population, the prevalence of chronic noncommunicable diseases, including osteoporosis in KSA, will increase if not remain steady, and uncorrected vitamin D deficiency being a risk factor for these diseases will play a major role in the progression of these diseases. It is worth to note that among these diseases, the emergence of increasing incidence of fibromyalgia or chronic muscle pain is mostly experienced by Saudi women, which showed significant improvement after treatment of high-dose vitamin D [52, 53], and reversal of metabolic syndrome manifestations among Saudi adults by mere increased sun exposure [54]. Several intervention studies are further required for the rest of the nonskeletal diseases where vitamin D is involved to determine whether vitamin D status correction will provide major beneficial effect.

Diseases	References
Chronic low back pain	[29]
Fibromyalgia	[30, 31]
Hyperparathyroidism	[32, 33]
Obesity and the metabolic syndrome	[21, 23, 24, 34, 35]
Osteoporosis/osteopenia/osteomalacia	[20–22, 36–44]
Sickle cell disease	[45, 46]
Systemic lupus erythematosus	[25]
Type-1 diabetes mellitus	[47–49]
Type-2 diabetes mellitus	[39, 50, 51]

http://dx.doi.org/10.5772/63638

Table 1. List of major diseases associated with vitamin D deficiency in Saudi Arabia.

4. Treatment

The two commonly available forms of vitamin D supplements are ergocalciferol (vitamin D_2) and cholecaleiferol (vitamin D_3). Some, but not all, studies suggest that vitamin D_3 increase serum 25[OH]D more efficiently than does vitamin D_2 [55–57]. The best indicator of vitamin D status is 25-hydroxyvitamin D because it is the major circulating form of vitamin D; it reflects cutaneous and dietary intake [58]. A nonfasting sample taken at any time of the day is suitable for the measurement of 25-hydroxyvitamin D status. Although calcitriol 1,25 dihydroxycholecalciferol is the active form of vitamin D, it is not an appropriate indicator of vitamin D status. It is usually normal or even elevated in patients with vitamin D deficiency. Although reliable and consistent evaluation of serum 25[OH]D level remains an issue, reliable laboratories currently exist, and efforts are in progress to improve and standardize assays to enhance accuracy and reproducibility at other laboratories.

Adults with 25 OHD 50–75 nmol/L require treatment with 800 to 1000 IU of vitamin D_3 daily. This intake was hypothesized to increase the vitamin D status to 7 nmol/L over a three-month period, but still, many individuals might require higher doses. In malabsorptive states, oral dosing and treatment duration depend on the individual patient's vitamin D absorptive capacity. Mega doses of vitamin D (10,000 to 50,000 IU daily) may be essential for postgastrectomy patients or patients with malabsorption. In cases where such patients remain deficient/insufficient despite such doses, they should be treated with hydroxylated vitamin D metabolites (since they are more readily absorbed) or with sun or sun camp exposure. All patients should maintain a daily calcium intake of at least 1000 mg (for ages 31 to 50 years) to 1200 mg (>51 years old) per day [59].

Since vitamin D is a fat-soluble vitamin, there are concerns about toxicity from excessive supplementation. Widespread fortification of food and drink from the 1930s to 1950s in the United States and Europe led to reported cases of toxicity. Increased levels of vitamin also raise calcium levels. Most of the symptoms of vitamin D toxicity are secondary to hypercalcemia. Early symptoms include, but are not limited to, gastrointestinal disorders like anorexia, diarrhea, constipation, nausea, and vomiting. Other reported symptoms include bone pain, drowsiness, continuous headaches, irregular heartbeat, loss of appetite, muscle, and joint pain are other symptoms that are likely to appear within a few days or weeks; frequent urination, especially at night, excessive thirst, weakness, nervousness and itching, and kidney stones [60].

5. Conclusion

This chapter provides a glimpse on the essential knowledge about this micronutrient vitamin D, as it is one of the most clinically important nutritional deficiencies. It is by no means comprehensive but nevertheless equips the reader with vital information on vitamin D with special attention in the Middle East and Saudi Arabia.

Conflict of interest

The author declares no conflict of interest.

Author details

Naji J. Aljohani

Address all correspondence to: najij@hotmail.com

King Fahad Medical City, College of Medicine, King Saud bin Abdulaziz University for Health Sciences, Riyadh, Saudi Arabia, and Prince Mutaib Chair for Biomarkers of Osteoporosis, Biochemistry Department, College of Science, King Saud University, Riyadh, Saudi Arabia

References

[1] DeLuca H. Overview of general physiological tentures and function of vitamin D. Am J Clin Nutr 2004;80(6 suppl.):16895–965.

[2] Mawer EB, Hayes ME, Hays SE, et al. Constitutive synthesis of 1,25 dihydroxy vitamin D_3 by a human small cell lung cancer cell line. J Clin Endocrinol Metab 1994;79:554–60.

[3] Schwartz GG, Whitlutch LW, Chen TC, Lokeshwar BL, Holick MF. Human prostate cells synthesize 1,25 dihydroxyvitamin D_3. Cancer Epidemiol Biomarker Prev 1998;7:391–95.

[4] Holick MF. Sunlight, vitamin D and health: a D-lightful story. In: The Norwegian Academy of Science and Letters, 2008. pp. 147–166.

[5] Pilz S, Tomaschitz A, Ritz E, Pieber TR. Vitamin D status and arterial hypertension: a systematic review. Nat Rev Cardiol 2009;6(10):621–30.

[6] Lips P. Vitamin D physiology. Prog Biophys Mol Biol 2006;92:4–8.

[7] Tai T, Need A, Horowitz M, Chapman I. Vitamin D, glucose, insulin, and insulin sensitivity. Nutrition 2008;24:279–85.

[8] Holick MF, Binkley NC, Bischoff-Ferrai HA, et al. Evaluation, treatment, and prevention of vitamin D deficiency: an endocrine society clinical practice guideline. J Clin Endocrinol Metab 2011;96(7):1911–30.

[9] Holick MF, Chen TC. Vitamin D deficiency: a worldwide problem with health consequences. Am J Clin Nutr 2008;87(4):1080S–6S.

[10] Hilger J, Friedel A, Herr R, et al. A systematic review of vitamin D status in populations worldwide. Br J Nutr 2014;111(1):23–45.

[11] Haq A, Svobodova J, Imran S, et al. Vitamin D deficiency: a single center analysis of patients from 136 countries. J Steroid Biochem Mol Biol 2016 [Epub ahead of print].

[12] Maalouf G, Gannage-Yared MH, Ezzedine J, et al. Middle East and North Africa consensus on osteoporosis. J Muskuloskelet Neuronal Interact 2007;7(2):131–43.

[13] Sedrani SH, Elidrissy AW, El Arabi KM. Low 25-hydroxyvitamin D and normal serum calcium concentrations in Saudi Arabia: Riyadh region. Ann Nutr Metab 1984;28(3): 181–5.

[14] Serenius F, Elidrissy AT, Dandona P. Vitamin D nutrition in pregnant women at term and in newly born babies in Saudi Arabia. J Clin Pathol 1984;37(4):444–7.

[15] Taha SA, Dost SM, Sedrani SH. 25-hydroxyvitamin D and total calcium: extraordinarily low plasma concentrations in Saudi mothers and their neonates. Pediatr Res 1984;18(8):739–41.

[16] Fonseca V, Tongia R, el-Hazmi M, Abu-Aisha H. Exposure to sunlight and vitamin D deficiency in Saudi Arabian women. Postgrad Med J 1984;60(707):589–91.

[17] Al-Arabi K, Elidrissy A, Sedrani S. Is avoidance of sunlight a cause of fractures of the femoral neck in elderly Saudis? Trop Georg Med 1984;36(3):273–9.

[18] Al-Mahroos F. Diabetes in the Arabian Peninsula. Ann Saudi Med 2000;20(2):111–2.

[19] Al-Daghri NM, Al-Attas OS, Alokail MS, et al. Diabetes mellitus type 2 and other chronic non-communicable diseases in the central region, Saudi Arabia (Riyadh Cohort 2): a decade of an epidemic. BMC Med 2011;9(1):76.

[20] Ardawi MS, Sibiany AM, Bakhsh TM, et al. High prevalence of vitamin D deficiency among healthy Saudi Arabian men: relationship to bone mineral density, parathyroid hormone, bone turnover markers, and lifestyle factors. Osteoporos Int 2012;23(2): 675–86.

[21] Ardawi MS, Qari MH, Rouzzi AA, et al. Vitamin D status in relation to obesity, bone mineral density, bone turnover markers and vitamin D receptor genotypes in healthy Saudi pre- and postmenopausal women. Osteoporos Int 2011;22(2):463–75.

[22] Sadat-Ali M, AlElq A, Al-Turki H, et al. Vitamin D levels in healthy men in eastern Saudi Arabia. Ann Saudi Med J 2009;29(5):378–82.

[23] Al-Daghri NM, Al-Attas OS, Al-Okail MS, Alkharfy KM, Al-Yousef MA, Nadhrah HM, Sabico SB, Chrousos GP. Severe hypovitaminosis D is widespread and more common in non-diabetics than diabetics in Saudi adults. Saudi Med J 2010;31(7):775–80.

[24] Al-Daghri NM, Al-Attas OS, Alokail MS, Alkharfy KM, Al-Othman A, Draz HM, Yakout SM, Al-Saleh Y, Al-Yousef M, Sabico S, Clerici M, Chrousos GP. Hypovitaminosis D associations with adverse metabolic parameters are accentuated in patients with diabetes mellitus type 2: a BMI-independent role of adiponectin? J Endocrinol Invest 2013;36(1):1–6.

[25] Damanhouri LH. Vitamin D deficiency in Saudi patients with systemic lupus erythematosus. Saudi Med J 2009;30(10):1291–5.

[26] Cannell JJ, Hollis BW. Use of vitamin D in clinical practice. Altern Med Rev 2008;13(1): 6–20.

[27] Chiu K, Chu A, Go V, Saad M. Hypovitaminosis D is associated with insulin resistance and β cell dysfunction. Am J Clin Nutr 2004;79:820–5.

[28] Chonchol M, Scragg R. 25-Hydroxivitamin D, insulin resistance, and kidney function in the Third National Health and Nutrition Examination Survey. Kidney Int 2006;71(2): 134–9.

[29] Al-Faraj S, Al Mutairi K. Vitamin D deficiency and chronic low back pain in Saudi Arabia. Spine (Phila Pa 1976) 2003;28(2):177–9.

[30] Fouda MA. Primary hyperparathyroidism and vitamin D deficiency: a combination still encountered in Asian countries. Ann Saudi Med 1999;19(5):455–8.

[31] Raef H, Ingemansson S, Sobhi S, et al. The effect of vitamin D status on the severity of bone disease and on the other features of primary hyperparathyroidism (pHPT) in a vitamin D deficient region. J Endocrinol Invest 2004;27(9):807–12.

[32] Ahmed M, Almahfouz A, Alarifi A, et al. Hyperparathyroidism secondary to vitamin D deficiency. Clin Nucl Med 2003;28(5):413–5.

[33] Ahmed M, Faraz HA, Almahfouz A, et al. A case of vitamin D deficiency masquerading as occult malignancy. Ann Saudi Med 2006;26(3):231–6.

[34] Al-Elq AH, Sadat-Ali M, et al. Is there a relationship between body mass index and serum vitamin D levels? Saudi Med J 2009;30(12):1542–6.

[35] Al-Sultan Al, Amin TT, Abou-Seif MA, et al. Vitamin D, parathyroid hormone levels and insulin sensitivity among obese young adult Saudis. Eur Rev Med Pharmacol Sci 2011;15(2):135–47.

[36] Ghannam NN, Hammani MM, Bakheet SM, et al. Bone mineral density of the spine and femur in healthy Saudi females: relation to vitamin D status, pregnancy, and lactation. Calcif Tissue Int 1999;65(1):23–8.

[37] Al-Jurayyan NA, El-Desouki ME, A-Herbish AS, et al. Nutritional rickets and osteomalacia in school children and adolescents. Saudi Med J 2002;23(2):182–5.

[38] El-Desouki MI, Othman SM, Fouda MA. Bone mineral density and bone scintigraphy in adult Saudi female patients with osteomalacia. Saudi Med J 2004;25(3):355–8.

[39] Al-Maatouq MA, El-Desouki MI, Othman SA, et al. Prevalence of osteoporosis among postmenopausal females with diabetes mellitus. Saudi Med J 2004;25(10):1423–7.

[40] Al-Osail AM, Sadat-Ali M, Al-Elq AH, et al. Glucocorticoid-related osteoporotic fractures. Singapore Med J 2010;51(12):948–51.

[41] Raef H, Al-Bugami M, Balharith S, et al. Updated recommendations for the diagnosis and management of osteoporosis: a local perspective. Ann Saudi Med 2011;31(2):111–28.

[42] Sadat-Ali M, El Elq AH, Al-Turki HA, et al. Influence of vitamin D levels on bone mineral density and osteoporosis. Ann Saudi Med 2011;31(6):602–8.

[43] Rouzi AA, Al-Sibiani SA, Al-Senani NS, et al. Independent predictors of all osteoporosis-related fractures among healthy Saudi postmenopausal women: the CEOR study. Bone 2012;50(3):713–22.

[44] Alissa EM, Qadi SG, Alhujaili NA, Alshehri AM, Ferns GA. Effect of diet and lifestyle factors on bone health in postmenopausal women. J Bone Miner Metab 2011;29(6): 725–35.

[45] Mohammed S, Addae S, Suleiman S, et al. Serum calcium, parathyroid hormone and vitamin D status in children and young adults with sickle cell disease. Ann Clin Biochem 1993;30(Pt 1):45–51.

[46] Sadat Ali-M, Al-Elq A, Al-Turki H, et al. Vitamin D level among patients with sickle cell anemia and its influence on bone mass. Am J Hematol 2011;86(6):506–7.

[47] Abdullah MA. Epidemiology of type 1 diabetes mellitus among Arab children. Saudi Med J 2005;26(6):911–7.

[48] Aljabri KS, Bokhari SA, Khan MJ. Glycemic changes after vitamin D supplementation in patients with type 1 diabetes mellitus and vitamin D deficiency. Ann Saudi Med 2010;30(6):454–8.

[49] Bin-Abbas BS, Jabari MA, Issa SD, Al-Fares AH, Al-Muhsen S. Vitamin D levels in Saudi children with type 1 diabetes. Saudi Med J 2011;32(6):589–92.

[50] Al-Shahwan MA, Al-Othman AM, Al-Daghri NM. Effects of 12-month, 2000IU/day vitamin D supplementation on treatment naïve and vitamin D deficient Saudi type 2 diabetic patients. Saudi Med J 2015;36(12):1432–8.

[51] Al-Daghri NM, Al-Saleh Y, Aljohani N, et al. Vitamin D deficiency and cardiometabolic risks: a juxtaposition of Arab adolescents and adults. PLoS One 2015;10(7):e0131315.

[52] Matthana MH. The relation between vitamin D deficiency and fibromyalgia syndrome in women. Saudi Med J 2011;32(9):925–9.

[53] Abokrysha NT. Vitamin D deficiency in women with fibromyalgia in Saudi Arabia. Pain Med 2012;13(3):452–8.

[54] Al-Daghri NM, Alkharfy KM, Al-Saleh Y, et al. Modest reversal of metabolic syndrome manifestations with vitamin D status correction: a 12 month prospective study. Metabolism 2012;61(5):661–6.

[55] Tang HM, Cole DE, Rubin LA, et al. Evidence that vitamin D_3 increases serum 25-hydroxyvitamin D more efficiently than does vitamin D_2. AM J Clin Nutr 1998;68:854.

[56] Armas LA, Hollis BW, Heaney RD. Vitamin D_2 is much less effective than vitamin D_3 in humans. J Clin Endocrinol Metab 2004;89:5387.

[57] Holick MF, Biancuzzo RM, Chen TC, et al. Vitamin D_2 is as effective as vitamin D_3 in maintaining circulating concentrations of 25-hydroxyvitamin D. J Clin Endocrinol Metab 2008;93:677–81.

[58] Institute of Medicine. Dietary reference intakes for calcium, phosphorus, magnesium, vitamin D, and fluoride. Washington, DC: National Academy Press; 1997.

[59] Standing Committee on the Scientific Evaluation of Dietary Reference Intakes. Fa NBIOM dietary reference intakes for calcium, phosphorus, magnesium, vitamin D, and fluoride. Washington, DC: National Academy Press; 1997.

[60] Schwalfenberg G. Not enough vitamin D: health consequences for Canadians. Can Fam Phys 2007;53(5):841–54.

Chapter 10

Cancer Treatment and Nutritional Deficiencies

Janet Schloss

Additional information is available at the end of the chapter

http://dx.doi.org/10.5772/63395

Abstract

Increasing cancer incidence and improved survival rates have seen the number of cancer survivors increase exponentially throughout the last few decades. As a consequence of this, cancer survivors may experience a number of permanent side effects from their cancer or the treatment. Traditionally, patient follow-up has been undertaken by oncological specialists with a major focus on possible cancer reoccurrence; however, this fails to identify or adequately address many patients' concerns regarding post-cancer treatment. For a majority of patients, nutrition during treatment and post-cancer diagnosis and treatment is an area they can control and change for their own health and well-being. The following chapter addresses nutrient deficiencies associated with certain cancers, chemotherapy agents, radiation and surgical procedures. Potential treatment protocols for different oncological stages post diagnosis are explored and conditions that may induce nutrient deficiencies and how they can be treated or decreased are explained.

Keywords: chemotherapy, radiation, cancer survivorship, nutrient deficiencies, well-being

1. Introduction

Increasing cancer incidence and improved survival rates have seen the number of cancer survivor's increase exponentially throughout the last few decades. As a consequence of this, cancer survivors may experience a number of permanent side effects from their cancer or the treatment [1]. Traditionally, patient follow-up has been undertaken by oncological specialists with a major focus on possible cancer reoccurrence; however, this fails to identify or adequately address many patients' concerns regarding post-cancer treatment. For a majority of patients, nutrition during treatment and post-cancer diagnosis and treatment is an area they can control and change for their own health and well-being.

However, Zhang (2015) [2] published a study indicating that cancer survivors are usually motivated to improve their health but were found to have suboptimal diets. She examined the dietary intake of 1533 cancer survivors and 3075 individuals who had never had cancer. The researcher estimated the quality of the diets using the Health Eating Index, which is based on the United States government's 2010 Dietary Guidelines for Americans. The scores ranged from 0 to 100, with 0 indicating no adherence to 100 which is total adherence. After adjusting for age, sex and ethnicity, Zhang found that the group who had not had cancer had an average index of 48.3 and the cancer survivors on average indexed 47.2. It was found that cancer survivors in general from this population ate less fibre, more empty calories and more refined sugars and fats. In addition, they examined patients who had different types of cancer and found that those who had breast cancer had the healthier diets and those who had lung cancer had the worst diets. It was also identified that cancer treatment may cause people to have specific food cravings, or change the way food tastes. This may influence the food choices they make post treatment.

Nutritional deficiencies in people with cancer who are undergoing traditional oncology treatment are a critical component for the health and survival of patients with or after cancer diagnosis. To date, a majority of research and nutritional screening has focused on malnutrition and weight loss in relation to nutritional deficiencies. This nutritional assessment is essential for the diagnosis of nutritional compromise as nutritional deterioration has been found to be associated with adverse outcomes in terms of cancer prognosis such as response rate and survival [3]. The nutritional screening and identification for malnutrition has been well documented. However, this screening omits the general patient undergoing treatment for cancer who is not elderly, malnourished or losing weight. Patients who have lung, oesophagus, stomach, colon, rectum, liver and pancreas cancer have been found to be at greatest risk of weight loss and malnutrition [3].

However, breast and prostate cancer, which are two of the most common cancers, have been found to be associated with weight gain, not weight loss [4]. To date, nutritional screening of patients undergoing adjuvant or neo-adjuvant chemotherapy has not been conducted to ascertain nutritional status. Research into possible nutritional insufficiencies may provide an insight to assist clinicians in aiding patients to thrive with or after cancer. Moreover, individual research has identified a number of nutritional deficiencies that can occur from certain chemotherapeutic agents, radiation and surgery. Combining the research that has been conducted for people with cancer or post cancer may provide the information necessary for clinicians and patients post diagnosis and treatment to live a healthy, balanced life based on nutrient sufficiency, not deficiency.

http://dx.doi.org/10.5772/63395

2. Background

Nutritional therapy for cancer requires a greater understanding of nutritional biochemistry, interactions as well as patients' expectations and disease impact. Nutritional analysis and early nutritional interventions (diet counselling, oral supplementation, enteral or total parenteral nutrition) may reduce, prevent or even reverse poor nutritional status, improve performance status and consequently affect their quality of life (QoL) [5]. The nutritional intervention may also depend on the type of cancer treatment, either curative or palliative. A nutritional intervention for a curative cancer treatment can have an additional role which is to increase the tolerance and response to the oncology treatment, decrease complications, reduce morbidity by optimizing the balance between energy expenditure and food intake, and decrease the possible risk of metastasis, whereas nutritional interventions for palliative care is aimed at improving the patients' QoL, controlling symptoms including vomiting, nausea, constipation and pain related to food intake [5].

Understanding the biochemistry associated with a patient who has a solid tumour versus a patient who is tumour-free post surgery/treatment is important for nutritional assessment. Cancer can have a major impact on a patient's physicality and psychological well-being. For example, proteolysis and lipolysis are accelerated while muscle protein synthesis is depressed in a person with a solid tumour. In addition, carbohydrate metabolism is modified by tumour growth such as an increased hepatic glucose production and Cori cycle activity and a reduction in insulin sensitivity in peripheral tissues. This results in a loss of lean body mass and fat tissue, causing an increase in energy expenditure and resulting in wasting [6, 7]. This type of cancer-related weight loss is different from simple starvation whereby normal refeeding can restore normal nutritional status. These tumour-associated metabolic abnormalities can frequently prevent the restoration of muscle mass and lead to cachexia due to complex interactions between pro-inflammatory cytokines and the host metabolism [8–10].

In addition to the effects of having a tumour, oncological treatments such as surgery, chemotherapy and radiotherapy can cause side effects and physiological changes that can affect food intake and nutritional status [11–14]. Moreover, the stress response from the treatment can have an effect on nutritional status and body composition. The changes in glucose metabolism, loss of muscle mass and increased fat distribution during chemotherapy also affect energy expenditure [15, 16]. In addition, the fatigue and nutritional status will vary depending on the patient who is assigned curative or palliative treatment.

A study conducted in 2015 on breast cancer patients analysed weight gain during adjuvant chemotherapy and survival. It was found that weight gain (between 1 and 12 kg) had a negative impact on both disease-free and overall survival rates [17]. Currently, the cause of weight gain during chemotherapy has not been revealed.

Individual nutrient deficiencies or insufficiencies can also occur during treatment. An example of this is vitamin D3. A systematic review published in 2013 found that 31% of cancer patients undergoing treatment were vitamin D3 deficient and 61% had insufficient levels [18]. The following chapter will investigate evidence-based research on nutrient deficiencies and insufficiencies during different phases of cancer treatment, stages and side effects of treatment.

3. Tumour-induced effects on nutritional status

The majority of research on tumour-induced effects on nutritional status is focused on cachexia or weight loss. Research has found that the progressive nutritional deterioration displayed in cachexia is different from starvation and is the result of the tumour burden on the body. The increased proteolysis and lipolysis is due to possible biochemical reactions in the body such as pro-inflammatory cytokine activation or specific molecules released by the tumour itself [19]. The proteolysis that is found in cachexia has also been found in cancer growth and can occur in individuals with a solid tumour who are not cachexic. The neoplasm or cancer growth can compromise the normal biochemical mechanisms that regulate muscle homeostasis, which results in the loss of muscle mass, functional impairment and compromised metabolism. The end result of this tumour-induced condition is enhanced muscle protein breakdown and amino acid release that sustains liver gluconeogenesis and tissue protein synthesis [20].

Research on individual nutrient deficiencies or insufficiencies has not been completed to date. It is uncertain at this stage if solid tumours cause nutrient deficiencies or nutrient insufficiencies. Further research is required to ascertain individual nutrient status of patients with cancer.

4. Nutritional implications from cancer therapies

Patients undergoing cancer treatment have been found to frequently experience malnutrition. The nutritional status of cancer patients varies depending on the treatment, the type of cancer and the ability to eat. A study on Indian patients published in 2015 investigated 57 cancer patients and evaluated them during treatment using a Patient-Generated Subjective Global Assessment (PG-SGA). The results found that 15.8% (9/57) were well nourished, 31.6% (18/57) were moderately or suspected of being malnourished and 52.6% (30/56) were found to be severely malnourished [21]. The researchers found that the highest malnutrition was in lip/oral cancer patients (33.3%) and that the prevalence of malnutrition was highest in patients during treatment (84.2%) [21].

Therefore, although not all nutrients have been researched to identify specific nutritional deficiencies or insufficiencies, it is highly likely that patients undergoing cancer treatment would have certain nutrient deficiencies or insufficiencies. These would vary just as patient responses to treatment vary as well.

4.1. Surgery

4.1.1. Head and neck cancers

Surgery for head and neck cancers includes tumours inside the sinuses, nose, mouth, salivary glands and down the throat including oesophageal cancer (Australian Cancer Research Foundation (ACRF)). The greatest impact on nutritional status from surgery for head and neck cancers is dysphasia (difficulty swallowing, approximately 14.7%) [22]. This impacts the patient's ability to eat and therefore nutrient intake. Research on specific nutrient deficiencies due to dysphasia has not occurred to date. Further research in this area is required.

4.1.2. Gastrointestinal cancers

The gastrointestinal cancers involve surgery for stomach (gastric), bowel (colorectal), liver, oesophageal in some cases, pancreatic, anal, bile duct, gastrointestinal carcinoid, gallbladder and small intestinal cancers (ACRF). Depending on the cancer, location, staging and possible metastasis will depend on the implications on nutritional status.

A study published in 2016 investigated lean body mass after gastrointestinal surgery [23]. The loss of lean body mass has been found to decrease the compliance of adjuvant chemotherapy particularly in patients undergoing gastrectomy for gastric cancer. The researchers examined 485 patients. They found that the median loss of lean body mass was 4.7%. In 225 patients (46.4%), a lean body mass of 5% or more occurred. A statistical significance was found using both uni- and multivariate logistic analysis for severe lean body mass loss due to surgical complications including infection or fasting (odds ratio (OR) = 3.576; $p = 0.001$), total gastrectomy (OR: 2.522; $p = 0.0001$) and gender (OR: 1.929; $p = 0.001$) [23].

Hence, the identification of nutritional intervention requirements of patients undergoing surgery for gastrointestinal cancer is required. This is an important factor and could impact on patient adjuvant treatment compliance and possible survival post surgery.

4.1.3. General surgery considerations

All surgical interventions for cancer will have some form of nutritional impact on patients. Individual assessment of patients prior to and post surgery is important for patient health, compliance and health/well-being through treatment and post treatment.

Considering that a large percentage of cancer patients undergo surgery for a biopsy or to remove tumours, lymph nodes or de-bulking a neoplasm, the human body requires support for both minor and major surgeries. The body is an amazing machine when supported correctly. The main nutritional support required is based on decreasing inflammation, supporting the immune system and the body to fight infection.

Traditionally, it is suggested to avoid alcohol, tobacco, simple sugars, processed foods and recreational drugs prior to and post surgery [24, 25]. Smoking and hazardous drinking have been found to be the most common lifestyle risk factors that influence surgery complications [25]. In addition, avoiding nutrient supplementation that could increase the risk of bleeding such as fish oils, vitamin E, turmeric and herbs such as ginkgo should be stopped before 1 week.

Antibiotic use is common in surgery pre- or postoperatively [26, 27]. Prophylactic use of antibiotics has been to prevent the potential risk of infection postoperatively as pre- and perioperative antibiotics have been found to lower the infection rate [26]. To assist the recolonization of the microbiota, it is recommended to use pre- and probiotics [28].

Possible nutrient deficiencies pre- and postoperatively such as iron [29] need to be taken into consideration in addition to possible insufficiencies and nutrients to assist in healing such as vitamin C, zinc and amino acids such as proline and glycine [30, 31]. The prevalence of nutrient deficiencies postoperatively has been mainly focused on bariatric patients rather than on cancer patients [32]. However, nutrients found to be deficient in these patients may be correlated to some cancer patients as a high percentage of patients with cancer have been found to have a higher body mass index (BMI) [33]. Therefore, nutrients such as vitamin D, which has been found to be deficient in approximately 57% of patients, vitamin B12, iron and folate, are best to be monitored pre- and postoperatively [32].

Hence, nutritional screening, management and support pre- and postoperatively assist the patient in chance for compliance through further interventional treatments in addition to survival.

4.2. Chemotherapy and immunotherapy

4.2.1. Identify certain nutrient deficiencies from chemotherapy administration

There are a large number of chemotherapy agents now on the market and are all divided into groups depending on their mechanism of action. Chemotherapy is often an effective treatment; however, each agent can cause particular side effects that can affect the person's health and well-being. Many of the new drugs now available do not cause the same severity of side effects and the new development in conventional medicine has helped to manage and reduce the main side effects of nausea, vomiting and leucopenia [34, 35].

Nutrient deficiencies that can occur from chemotherapy have limited research. A common side effect is chemotherapy-induced anaemia; however, this is not caused by low iron levels or deficiency. This side effect is due to the chemotherapy agent's mechanism of action on the development of red blood cells. Supplemental iron has been effective for an iron deficiency but not for chemotherapy-induced anaemia. Too much iron may promote tumour growth or worsen chemotherapy side effects. Therefore, iron supplementation should only be recommended if there is a diagnosed iron deficiency confirmed by pathology tests.

Vitamin B12 has been found to be deficient in certain individuals after chemotherapy [36]. A case study was presented in which a patient in a clinical trial for chemotherapy-induced peripheral neuropathy was found to be deficient in vitamin B12 post chemotherapy. This woman had normal vitamin B12 blood parameters pre-chemotherapy administration and again upon intramuscular vitamin B12 injection and supplemental vitamin B12 and 6 months after supplementation. Although this represents only one individual, it is possible that certain individuals may develop vitamin B12 deficiencies during chemotherapy, which may induce more severe presentation of other chemotherapy-induced side effects.

Hereditary disorders that cause haemolytic anaemias have also been found to induce a vitamin B12 deficiency, which require lifelong vitamin B12 administration [37]. These conditions need to be identified prior to chemotherapy administration to ensure that the patient is not in a deficient state. Another consideration is the use of protein pump inhibitors (PPI) and histamine H2-receptor antagonists as an association has been found with their use and a vitamin B12 deficiency [38]. PPIs are used during chemotherapy to assist with reflux and could have an impact on vitamin B12 absorption. In addition, metformin is another drug that has been found to decrease vitamin B12 and in combination with either histamine H2-receptor antagonists or PPIs, neuropathy due to vitamin B12 depletion has been found [39].

Another vitamin that has been found to be deficient during chemotherapy is vitamin D3. Teleni et al. in 2013 conducted a meta-analysis on vitamin D3 status in cancer patients [18]. They found that 31% of patients undergoing active treatment were deficient in vitamin D3 and 67% had insufficient levels. These findings and the awareness, impact and importance of vitamin D3 in the medical fraternity have now seen it being one nutrient that has been commonly prescribed to cancer patients undergoing treatment.

The main mineral that has been found to be deficient in patients undergoing chemotherapy such as cetuximab is magnesium [40]. Hypomagnesaemia has also been found in patients on PPIs particularly in combination with diuretics [41], which are common medications used in conjunction with chemotherapy agents. It is important to monitor magnesium levels in patients and potential oral supplementation may be required.

Research on nutrients to assist side effects from chemotherapy has continued; however, nutrients that are depleted during chemotherapy are still required. Potential nutrient deficiencies rather than macronutrient depletion may play an important role in patient mortality or morbidity. Further research is required to ascertain possible insufficiencies and deficiencies that could contribute to poor health and well-being of patients diagnosed with cancer and undergoing chemotherapy.

4.3. Radiation

Radiation, similar to chemotherapy, is considered to be an effective treatment against actively dividing cells. According to the American Cancer Society, more than 50% of all cancer patients undergo radiotherapy (www.cancer.org). Nutritional impact from radiation depends on where the person is receiving radiation. Head and neck cancers, lung cancer and gastrointestinal cancers have been found to have the greatest nutritional impact on cancer patients. The nutritional status of patients undergoing radiation therapy has been assessed, with specific nutritional indicators measured. One particular study focused on chemoradiotherapy on nasopharyngeal cancer. They found that after radiotherapy, 20.2% of patients had more than 10% weight loss. Statistically significant ($p = 0.05$) risk factors for poor nutritional status included old age, females, late stage of the disease, depression, high side effects and moderate nutritional status prior to radiotherapy [42]. It is advised that patients undergoing radiotherapy, particularly head and neck, gastrointestinal and lung cancer patients, be nutritionally assessed and intervention commenced to prevent malnutrition during treatment.

Individual nutrient screening of patients undergoing radiation is extremely limited. The main nutritional research on radiation is based on the prevention of malnutrition and weight loss, particularly for head and neck cancers. The importance of early nutritional management and intervention has been stipulated and implementation in hospitals has been encouraged [43]. Further research into individual nutrient deficiencies and insufficiencies during radiotherapy may also contribute to the health and outcome of cancer patients.

5. Nutritional screening

Effective nutritional screening, implementation of nutritional care plans and support are essential components for cancer patients. The screening and early detection of malnutrition is considered crucial in identifying patients at nutritional risk. A high prevalence of malnutrition has been identified in hospitalized cancer patients undergoing treatment, for example, colorectal cancer [44].

5.1. Current screening and assessment tools

Currently, there are a number of nutritional assessment tools used in clinical practice for cancer patients. The accuracy of diagnostic tools is based on sensitivity, specificity and positive- and negative-predictive values calculated on the likelihood that a given test result would be expected when the target condition is present compared with the likelihood of the same result if the condition was absent [44].

Tables 1 and **2** evaluate the nutritional tools available. The information has been obtained from the Queensland Government of Australia who conducted and published a malnutrition screening and assessment tool comparison in addition to a validated nutrition assessment tool comparison [45]. The screening tools evaluated used the parameters such as recent weight loss, poor intake/appetite and body weight measurements. It was found that all tools evaluated generally performed well. Choosing the correct nutritional screening tool will depend on various aspects such as complexity, sensitivity to that population group, who will be performing the screening, what actions will be undertaken and how the outcomes will be incorporated into the current facility procedures [45].

Name author, year	Setting and patient population	Nutrition assessment parameters	Rationale/clarification
Subjective Global Assessment (SGA) 1987 [46]	Setting: Acute [47–49] Rehab [50] community [51] Residential aged care [52] Patient group: Surgery [47] Geriatric [50–53] Oncology [48] Renal [49]	Medical history (weight, intake, GI symptoms, functional capacity) and physical examination Categories: 1. SGA A (well nourished) 2. SGA B (mild-moderate malnutrition) 3. SGA C (severe malnutrition	Requires training Easy to administer Good intra- and inter-rater reliability Patent-Generated
Subjective Global Assessment (PG-SGA) Ottery, F. 2005 [54]	Setting: Acute [55–57] Patient group: Oncology [55]	Medical history (weight, intake, symptoms, functional capacity, metabolic demand) and physical examination	Numerical score assists in monitoring changes in nutritional status

Name author, year	Setting and patient population	Nutrition assessment parameters	Rationale/clarification
http://pt-global.org/	Renal [56] Stroke [57]	Categories: SGA categories (A, B or C) as well as providing a numerical score for triaging. Global categories should be assessed as per SGA.	Easy to administer Scoring can be confusing requires training Patients can complete the first half of the tool by themselves
Mini-Nutritional Assessment (MNA) Guigoz Y et al. 1994 [58] http://www.mna-elderly.com/	Setting: Acute [58] Community [58] Rehab [58] Long-term care [58] Patient group: Geriatric [58]	Screening and assessment component includes diet history, anthropometry (weight history, height, MAC, CC), medical and functional status. Assessed based on numerical score as: - no nutritional risk - at risk of malnutrition or - malnourished	Lengthy Low specificity for screening section of tool in acute populations Can be difficult to obtain anthropometric data in this patient group

Table 1. Validated nutrition assessment tools: comparison guide [45].

Name author, year, country	Patient population	Nutrition screening parameters	Criteria for risk of malnutrition	When/by whom	Reliability established	Validity established
Malnutrition Screening Tool (MST) [59] Ferguson et al. (1999) Australia	Acute adults: inpatients and outpatients [59, 60] Elderly [61] Residential aged-care facilities [61]	Recent weight loss Recent poor intake	Score 0–1 for recent intake Score 0–4 for recent weight loss Total score: > 2 = at risk of malnutrition	Within 24 h of admission and weekly during admission. Medical, nursing, dietetic, admin staff; family, friends, patients themselves	Agreement by 2 Dieticians in 22/23 (96%) cases Kappa = 0.88 Agreement by a Dietician and Nutrition Assistant in 27/29 (93%) of cases Kappa = 0.84; and 31/32 (97%) of cases Kappa = 0.93	Compared with SGA and objective measures of nutrition assessment. Patients classified at high risk had longer length of stay. Sensitivity = 93% Specificity = 93%
Mini-Nutritional Assessment – Short Form (MNA-SF) [62] Rubenstein et	Elderly Best used in community, subacute or residential aged-care	Recent intake Recent weight loss Mobility Recent acute disease or psychological stress	Score 0–3 for each parameter Total score: <11 = at risk, continue with MNA	On admission and regularly not stated	Not reported	Compared to MNA and clinical nutritional status. Sensitivity = 97.9% Specificity = 100%

Name author, year, country	Patient population	Nutrition screening parameters	Criteria for risk of malnutrition	When/by whom	Reliability established	Validity established
al. (2001) United States	settings, rather than acute care [63]	Neuropsychological problems BMI				Diagnostic accuracy = 98.7% Compared with SGA in older inpatients Sensitivity = 100% Specificity = 52%2
Malnutrition Universal Screening Tool (MUST) [64] Malnutrition Advisory Group, BAPEN (2003) UK	Adults – acute and community	BMI Weight loss (%) Acute disease effect score	Score 0–3 for each parameter Total score: >2 = high risk 1 = medium risk 0 = low risk	Initial assessment and repeated regularly Able to be used by all staff	Internally consistent and reliable. Very good to excellent reproducibility Kappa = 0.8–1.0	Face validity, content validity, concurrent validity with other screening tools (MST and NRS) [65] Predicts mortality risk and increased length of stay and discharge destination in acute patients [66]
Nutrition Risk Screening (NRS-2002) [67] Kondrup et al. (2003) Denmark	Acute adult	% of recent weight loss % of recent poor intake BMI Severity of disease Elderly	Score 0–3 for each parameter Total score: >3 = start nutritional support	At admission and regularly during admission Medical and nursing staff	Good agreement between a Nurse, Dietician and Physician Kappa = 0.67	Retrospective and prospective analysis. Tool predicts higher likelihood of positive outcome from nutrition support and reduced length of stay among patients selected at risk by the screening tool and provided nutrition support.

Table 2. Comparison of malnutrition assessment and screening tools [45].

5.2. Nutritional deficiencies linked with specific cancers

5.2.1. Breast cancer

Breast cancer has been found to be the most frequently diagnosed cancer in women worldwide. It is estimated that 1.7 million cases and 521,000 deaths in 2012 were attributed to breast cancer and breast cancer alone accounts for 25% of all cancer cases and 15% of all cancer deaths among females [68]. There have been a number of different nutrient deficiencies or insufficiencies that have been attributed to an increased risk of breast cancer development. These include vitamin D3, iodine, folate, zinc, betacarotene and coenzyme Q10. **Table 3** shows the association of these nutrients and the risk of breast cancer.

Nutrient	Outcome
Coenzyme Q10	One study in 1998 investigated the role of coenzyme Q10 or ubiquinone in 200 women hospitalized for a biopsy and/or ablation of a breast tumour. They found that 80 patients (40%) with carcinomas and 120 patients (60%) with a non-malignant lesion had a coenzyme Q10 deficiency. There was also a correlation between the intensity of the deficiency and the prognosis of the breast cancer severity [69].
Folate	A lot of focus has been placed on the methylenetetrahydrofolate reductase (MTHFR) polymorphisms of late. A case-controlled study and pooled meta-analysis conducted in 2007 found that peri-menopausal ladies with the C677T polymorphism did have an increased risk of developing breast cancer [70].
Folate, zinc, betacarotene	A recent study in 2014 found that multiple genetic polymorphisms and/or deficiencies in folate, zinc and betacarotenes were associated with the triple negative breast cancer development, particularly in combination [71].
Iodine	Iodine was presented as a possible anti-proliferative agent for mammary glands in 2005 [72]. It has been found in both animal and human studies to exert a suppressive effect on the development and size of benign and cancer neoplasms [72]. As iodine stores in the thyroid and breast tissue, it exerts a protective action on the development of breast cancer. As hypothyroidism has been found to be high in breast cancer patients, it is proposed that low iodine levels may be considered a risk factor for breast cancer [73].
Selenium	In a meta-analysis conducted in 2014, an inverse relationship was found between selenium serum levels and the risk of breast cancer [74]. Therefore, maintaining selenium levels may decrease the risk of breast cancer for some women.
Vitamin D3	A vitamin D deficiency is highly prevalent among breast cancer females [75]. A vitamin D deficiency has been found in 99% of breast cancer females at diagnosis and approximately in 90% in healthy females [76]. Alcohol status and weight have an impact on vitamin D status and breast cancer risk [77].

Table 3. Nutrient deficiencies and breast cancer risk.

5.2.2. *Prostate cancer*

Nutrient	Outcome
Selenium	A systematic review and meta-analysis of selenium and prostate cancer found that the relationship between plasma/serum selenium and prostate cancer showed that the risk of developing prostate cancer decreased with increasing plasma/serum selenium levels (170 ng/mL) [78]. Further studies are required but there is a link between low selenium levels and prostate cancer risk.
Vitamin D3	Vitamin D3 (25(OH)D concentrations have been found to be inversely correlated with prostate cancer risk but not vitamin D–related polymorphisms or parathyroid hormone. This indicates that there is a possibility that low vitamin D3 blood pathology may pose a risk of prostate cancer risk [79]. No association has been found to vitamin D levels or vitamin D supplementation on prostate-specific antigen (PSA) levels [80]. It has been suggested that adding vitamin D supplementation might be an economical and safe way to possibly reduce the prostate cancer incidence and improve the cancer prognosis and outcome [81].
Vitamin E and trace minerals	As mentioned, a study on Nigerian prostate cancer males was conducted. This study showed that the levels of whole blood superoxide dismutase (SOD), vitamin E, serum selenium and zinc were significantly lower in prostate cancer patients. Therefore, the authors conclude that deficiencies in vitamin E, zinc and selenium may be risk factors for the development of prostate cancer [82].
Zinc	Human studies on zinc deficiencies and prostate cancer are limited. In vitro studies have found that a zinc deficiency does impact prostate cells and can compromise DNA integrity by impairing the function of zinc-containing proteins [83, 84]. One study conducted on Nigerian prostate cancer patients did find an association with a zinc deficiency and prostate cancer in addition to selenium and vitamin E deficiencies [82].

Table 4. Nutrient deficiencies linked with prostate cancer.

A majority of the population feel that prostate cancer is the most frequently diagnosed cancer in men worldwide, but in fact it is the second with 1.1 million new cases estimated to have occurred in 2012. However, it is the most frequently diagnosed cancer in men in developed countries. The incidence rates vary with the highest rates found in Australia/New Zealand, Northern America, Northern and Western Europe and some Caribbean nations. The lowest incidence rates are found in the Asian countries [68]. Nutrient deficiencies that have been studied and identified as potential risk factors include vitamin D3, selenium, zinc and vitamin E (**Table 4**).

5.2.3. *Colon cancer*

Colon or colorectal cancer is the third most commonly diagnosed cancer in males and second in females. It is estimated that 1.4 million cases and 693,000 deaths occurred in 2012 due to colorectal cancer. The highest incidence rates have been found in Australia/New Zealand, Europe and North America. The lowest incidence rates are found in Africa and South-Central Asia [68]. Nutrients that have been associated with an increased colon cancer risk include vitamin D3 and folate. Folic acid is controversial with a deficiency and if excess is linked with

http://dx.doi.org/10.5772/63395

colorectal cancer risk. In addition to specific nutrients, dietary factors are linked with colorectal cancer development as seen in **Table 5**.

Nutrient	Outcome
Fibre, low-fruit and -vegetable, high red and processed meat intake	Although not a specific nutrient, it has been well established that a diet low in fruits and vegetables, fibre and high in red and processed meat intake is a risk factor for colorectal cancer development [85–87].
Folic acid	Folic acid is a controversial nutrient for colorectal cancer. High levels have been associated with a reduced colorectal cancer risk; however, excessive folate levels may promote tumour progression [88]. These facts have prevented countries fortifying foods with folate due to the risk of colorectal cancer. Preventing a deficiency in folic acid is recommended as it is a risk factor for cancer development but monitoring levels to prevent excess is also recommended.
Selenium	Animal studies have found that a selenium deficiency can acerbate colitis and promote tumour development and progression in inflammatory carcinogenesis [89].
Vitamin D3	Vitamin D may protect and treat inflammatory bowel disease and assist colon cancer [90]. Vitamin D3 deficiency and insufficiency has been linked as a risk factor for colorectal cancer as found in observational studies in both human and experimental studies (animal and cell lines). The protection from vitamin D3 has been attributed its influence on cell proliferation, differentiation, apoptosis, DNA repair mechanism, inflammation and immune function [91]. A high prevalence of a vitamin D3 deficiency and insufficiency has been found in colorectal cancer patients [75].

Table 5. Nutrient deficiencies linked with colorectal cancer.

5.2.4. *Lung cancer*

Nutrient	Outcome
Selenium	Several epidemiological studies have shown an increased risk of lung cancer among adults with low blood levels of selenium; however, the results are inconsistent. One study conducted in the south-eastern United States found that there was a risk of lung cancer development in lower income and black Americans [92].
Vitamin A	Cigarette smoking has been directly associated with the development of lung cancer. It has been demonstrated that cigarette smoke significantly reduces retinoic acid in the lungs of rats and increases the formation of precancerous and cancerous lesions [93]. It has been found that this is attributed to two independent pathways, RARα- and RARβ-mediated pathways. Human studies are limited if a vitamin A deficiency increases the risk of lung cancer development if exposed to cigarette smoke.

Nutrient	Outcome
Vitamin D3	A high prevalence of low vitamin D3 has been found in lung cancer patients ranging from a mild deficiency to severe deficiencies [75].
Zinc	Human studies on zinc deficiency and lung cancer are limited. Cell culture work on human lung fibroblasts has found that a zinc deficiency can cause DNA instability and compromise its integrity and therefore may be important in the prevention of DNA damage and cancer [94].

Table 6. Nutrient deficiencies linked with lung cancer.

The popularity of breast and prostate cancer override the one cancer that is the most frequent cause of death among males in 2012 and is the leading cause of death in females in developed countries and second in less developed countries, lung cancer. The highest lung cancer incidence rates include Europe, Eastern Asia and Northern America and the lowest rates are in sub-Saharan Africa. Although smoking has high correlation with lung cancer development, a high prevalence of non-smoking individuals has been diagnosed with lung cancer. This high prevalence has been thought to reflect indoor air pollution, cooking fumes, exposure to occupational and environmental carcinogens such as asbestos, arsenic, radon and polycyclic aromatic hydrocarbons. Recently, outdoor pollution as also been attributed as a cause of lung cancer [68]. In addition, certain nutrients may also play a role in the development of lung cancer. These include vitamin D3, zinc, vitamin A and selenium as seen in **Table 6**.

6. Nutritional deficiencies and therapy for certain conditions linked with cancer treatment

Condition	Possible nutrient deficiency or insufficiency
Alteration of taste and smell	Zinc
Cachexia	Multiple nutrient deficiencies, protein, essential fats
Chemotherapy-induced peripheral neuropathy	Vitamin B12, vitamin B6, vitamin E, omega 3 fatty acid (DHA)
Dehydration	Water, electrolytes
Diarrhoea	Water, electrolytes, gut bacteria (lactobacillus, bifidus etc.)
Eczema/dermatitis	Essential fats, omega 3 fatty acids, vitamin E, vitamin D3, zinc, vitamin A
Hand and foot syndrome	Vitamin B6
Mucositis	Glutamine, vitamin A, zinc, glucosamine, vitamin C
Radiation-induced enteritis	Glutamine, vitamin A, zinc, glucosamine, vitamin C

Table 7. Potential nutritional deficiencies or insufficiencies for conditions linked with cancer treatment.

http://dx.doi.org/10.5772/63395

Research into nutrient deficiencies linked with certain conditions is limited. Certain nutrients have been found to be insufficient or deficient for certain conditions and may assist the patient in managing the situation. **Table 7** lists some conditions and possible nutrients, which could be found to be deficient or insufficient. It may be beneficial to consider the replacement of these nutrients for patients, or at least pathological or physical assessment to check the status.

7. Nutrition for patients who have had treatment for curative cancer

Curative cancer treatment normally occurs after surgery and can be intense. The impact on the nutritional status of the patient strongly depends on the tumour site, stage and progression of the cancer, the risks of the active treatment and the base nutritional status of the patient. For example, a patient undergoing concurrent chemotherapy and radiation for head and neck or lung cancer has a higher risk of malnutrition and impact on nutritional status than a patient undergoing adjunct chemotherapy for breast cancer.

Nutritional assessment and management should be started at the time of diagnosis and monitored throughout active treatment and afterwards. An ideal nutritional intervention and management commences with the initial evaluation of the patient's nutritional status through preliminary assessment tools and blood pathology tests. Regular re-evaluation is required throughout the treatment and post treatment until a good nutritional status is restored.

8. Nutrition in advanced cancer/palliative

Advanced cancer or palliative treatment is defined as patients who have metastatic cancer or are not responsive to curative treatment [5]. The life expectancy for these patients can vary from 1 month to many years. Therefore, nutritional assessment and intervention will depend on the stage of the cancer, the individual's current state, controlling the symptoms, maintaining an adequate hydration state and maintaining or restoring the patients 'well-being'.

Body weight will vary depending on the person as weight gain can occur due to lack of mobility and fatigue or weight loss/cachexia towards the end of life. Oedema and ascites from the tumour sites can also cause discomfort and impact digestive ability. Nutritional intake can also influence the QoL of the patient [5]. Constant re-evaluation and nutritional options are required as the patient's physical state changes. Consideration of nutrient intake, supplementation and nutritional fluid replacements are all important for each stage. Optimal nutritional status may not be restored in some cases; however, maintaining nutritional status for as long as possible has been found to be beneficial for the patient's well-being and QoL [5].

9. Summary

Current treatment for cancer is focused on survival, cure or pain management of the patient through active treatments such as surgery, chemotherapy, immunotherapy, radiation or hormone treatment. The nutritional status of patients generally is not a major consideration of primary health professionals unless malnutrition or weight loss is present, or the treatment may induce malnutrition. However, with the increasing number of cancer survivors, base nutritional status, nutritional assessment and support need to be extended to all cancer patients prior, during and post active cancer treatment. Nutritional screening and assessment needs to be considered an essential component of all aspects of cancer treatment.

This increased likelihood of individuals with cancer living longer after treatment has seen 'cancer survivorship' become a popular concept amongst organizations, hospitals, institutions and researchers within the field of oncology. A cancer 'survivor' is commonly defined as any person who has been diagnosed with cancer from the time of diagnosis through the balance of their life [95], although, many parties advocate for use of the term to relate to individuals who have had a previous cancer diagnosis and are now pursing life 'after active treatment' [95]. There are three distinct phases of cancer survivorship: *time of diagnosis to active treatment, the transition from active treatment to extended survival* and *long-term survival* [96]. In 2013, the American Society of Clinical Oncology (ASCO) released its assessment of survivorship care in adults [97] and conducted its first Inaugural Survivorship Symposium this year, 2016, in San Francisco.

One of the main focuses of cancer survivorships is diet, nutrition, exercise and long-term side-effect management. From definition, this starts from cancer diagnosis. Potential nutrient deficiencies or insufficiencies are areas that need further attention as well as their possible impact on side effects experienced by patients. Integration of nutritional assessment and intervention can be achieved through the current medical system and should be an important component of cancer patient-centred care.

10. Further directions

- Investigation into nutrient deficiencies in newly diagnosed cancer patients with emphasis on the type of cancer, social and economic status, gender and culture.
- Future trials and/or nutritional monitoring, assessment and intervention throughout cancer patient's active treatment and post treatment.
- Research into how nutritional deficiencies or insufficiencies may affect patient side effects to treatment.
- Research into potential nutrient deficiencies and insufficiencies as risk factors for cancer development and their mechanisms of action in cell impairment and cancer initiation and progression.

http://dx.doi.org/10.5772/63395

Author details

Janet Schloss

Address all correspondence to: janet.schloss@uqconnect.edu.au

1 Office of Research, Endeavour College of Natural Medicine, Brisbane, Australia

2 The School of Medicine, University of Queensland, Brisbane, Australia

References

[1] Aziz NM, Rowlands J. Trends and advances in cancer survivorship research: challenge and opportunity. Semin Radiat Oncol. 2003. 13:248–66.DOI:10.1016/S1053-4296(03)00024-9

[2] Zhang FF, Liu S, John EM, Must A, Demark-Wahnefried W. Diet quality of cancer survivors and noncancer individuals: results from a national survey. Cancer. 2015. 121:4212–21.DOI:10.1002/cncr.29488

[3] Capra S, Ferguson M, Ried K. Cancer: impact of nutrition intervention outcome—nutrition issues for patients. Nutrition. 2001. 17:769–72. DOI:10.1016/S0899-9007(01)00632-3

[4] Kwok A, Palermo C, Boltong A. Dietary experiences and support needs of women who gain weight following chemotherapy for breast cancer. Support Care Cancer. 2015. 23:1561–8.DOI:10.1007/s00520-014-2496-5

[5] Caroa MMM, Lavianob A, Pichard C. Nutritional intervention and quality of life in adult oncology patients. Clin Nutr. 2007. 26:289–301. DOI:10.1016/j.clnu.2007.01.005

[6] Delano MJ, Moldawer LL. The origins of cachexia in acute and chronic inflammatory diseases. Nutr Clin Pract. 2006. 21:68–81.DOI:10.1177/011542650602100168

[7] Laviano A, Meguid M, Inui A, Muscaritoli M, Rossi-Fanelli F. Therapy insight: cancer anorexia-cachexia syndrome – when all you can eat is yourself. Nat Clin Pract Oncol. 2005. 2:158–65.DOI:10.1038/ncponc0112

[8] Van Cutsem E, Arends J. The causes and consequences of cancer-associated malnutrition. Eur J Oncol Nurs. 2005. 9:S51–63.DOI:10.1016/j.ejon.2005.09.007

[9] Argiles M. Cancer-associated malnutrition. Eur J Oncol Nurs. 2005. 9(Supp 2):S39–50. DOI:10.1016/j.ejon.2005.09.006

[10] Fearon KC, Moses A. Cancer cachexia. Int J Cardiol. 2002. 85:73–81.

[11] Heys SD, Schofield AC, Wahle KW, Garcia-Caballero M. Nutrition and the surgical patient: triumphs and challenges. Surgeon. 2005. 3:139–44. DOI:10.1016/S1479-666X(05)80033-2

[12] Kiyama T, Mizutani T, Okuda T, Fujita I, Tokunaga A, Tajiri T, Barbul A. Postoperative changes in body composition after gastrectomy. J Gastrointest Surg. 2005. 9:313–9.DOI:10.1016/j.gassur.2004.11.008

[13] Bergkvist K, WengstromY. Symptom experiences during chemotherapy treatment–with focus on nausea and vomiting. Eur J Oncol Nurs. 2006. 10:21–9. DOI:10.1016/j.ejon.2005.03.007

[14] Guren MG, Tobiasseb LB, Trygg KU, Drevon CA, Dueland S. Dietary intake and nutritional indicators are transiently compromised during radiotherapy for rectal cancer. Eur J Clin Nutr. 2006. 60:113–9. DOI:10.1038/sj.ejcn.1602274

[15] Groenvold M. Health-related quality of life in early breast cancer. Dan Med Bull. 2010. 57:B4184.

[16] Antonella B, Rachna K, Martine E. Toxicity of treatment. Am J Clin Oncol. 2011. 34:292–6.

[17] Atalay C, Kucuk AI. The impact of weight gain during adjuvant chemotherapy on survival in breast cancer. Ulus Cerrahi Derg. 2015. 31:124–7.

[18] Teleni L, Baker J, Koczwara B, Kimlin MG, Walpole E, Tsai K, Isenring EA. Clinical outcomes of vitamin D deficiency and supplementation in cancer patients. Nutr Rev. 2013. 71:611–21.

[19] Cravo ML, Gloria LM, Claro I. Metabolic responses to tumour disease and progression: tumour–host interaction. Clin Nutr. 19:459–465.

[20] Sandri M. Protein breakdown in cancer cachexia. Semin Cell Dev Biol. 2015. DOI:10.1016/j.semcdb.2015.11.002

[21] Sharma D, Kannan R, Tapkire R, Nath S. Evaluation of nutritional status of cancer patients during treatment by patient-generated subjective global assessment: a hospital-based study. Asian Pac J Cancer Prev. 2015. 16:8173–6.

[22] Fujiki M, Sakuraba M, Miyamoto S, Hayashi R. Predictive factors of dysphagia after lateral and superior oropharyngeal reconstruction with free flap transfer. J Surg Oncol. 2016. 113:240–3.

[23] Aoyama T, Sato T, Segami K, Maezawa Y, Kano K, Kawabe T, Fujikawa H, Hayashi T, Yamada T, Tsuchida K, Yukawa N, Oshima T, Rino Y, Masuda M, Ogata T, Cho H, Yoshikawa T. Risk factors for the loss of lean body mass after gastrectomy for gastric cancer. Ann SurgOncol. 2016. DOI:10.1245/s10434-015-5080-4

[24] Zwissler B, Reither A. Preoperative abstinence from smoking. An outdated dogma in anaesthesia? Anaesthesist. 2005. 54:550–9.

[25] Tønnesen H, Nielson P, Lauritzen JB, Møller AM. Smoking and alcohol intervention before surgery: evidence for best practice. BJA. 102:297–306.

[26] Jones DJ, Bunn F, Fell-Syer SV. Prophaylactic antibiotics to prevent surgical site infection after breast cancer surgery. Cochran Database Syst Rev. 2014. 9:CD005360.

[27] Yany WB, Li CJ, LJ Li, Sheng SR, Qi SQ, Pan J. Postoperative infection bacteria and drug resistance in patients with oral and maxillofacial tumors. Shanghai Kou Qiang Yi Xue. 2015. 24:584–8.

[28] Andermann TM, Rezvani A, Bhatt AS. Microbiota manipulation with prebiotics and probiotics in patients undergoing stem cell transplantation. Curr Hemotol Malig Rep. 2016. DOI:10.1007/s11899-016-0302-9

[29] Clevenger B, Mallett SV, Klein AA, Richards T. Patient blood management to reduce surgical risk. Br J Surg. 2015. 102:1325–37.

[30] Moores J. Vitamin C: a wound healing perspective. Br J Community Nurs. 2013. S6:S8–11.

[31] Kurmis R, Greenwood J, Aromataris E. Trace element supplementation following severe burn injury: asystematic review and meta-analysis. J Burn Care Res. 2015. DOI: 10.1097/BCR.0000000000000259

[32] Toh SY, Zarshenas N, Jorgensen J. Prevalance of nutrient deficiencies in bariatric patients.Nutrition. 2009. 25(11–12):1150–6.

[33] Caan BJ, Kwan M, Hartzell G, Castillo A, Slattery ML, Sternfeld B, Weltzien E. Pre-diagnosis body mass index, post-diagnosis weight change, and prognosis among women with early stage breast cancer. Cancer Causes Control. 2008. 19:1319–1328.

[34] Chiu L, Chow R. Popovic M, Navari RM, Shumway NM, Chiu N, Lam H, Milakovic M, Pasetka M, Vuong S, Chow E, DeAngelis C. Efficacy of olanzapine for the prophylaxis and rescue of chemotherapy-induced nausea and vomiting (CINV): a systematic review and meta-analysis. Support Care Cancer. 2016. DOI:10.1007/s00520-016-3075-8

[35] Buchner A, Elsasser R, Bias P. A randomized, double-blind, active control, multicenter, dose-finding study of lipegfilgrastim (XM22) in breast cancer patients receiving myelosuppressive therapy. Breast Cancer Res Treat. 2014. 148:107–16.

[36] Schloss JM, Colisimo M, Airey C, Vitetta L. Chemotherapy-induced peripheral neuropathy (CIPN) and vitamin B12 deficiency. Support Care Cancer. 2015. 23:1843–50.

[37] Dietzfelbinger H, Hubmann M. Hemolytic anemias and vitamin B12 deficieny. Dtsch Med Wochenschr. 2015. 140:1302–10.

[38] Ruscin JM, Page RL 2nd, Valuck RJ. Vitamin B(12) deficiency associated with histamine(2)-receptor antagonists and a proton-pump inhibitor. Ann Pharmacother. 2002. 36:812–6.

[39] Zdilla MJ. Metformin with either histamine H2-receptor antagonists or proton pump inhibitors: apolypharmacy recipe for neuropathy via vitamin B12 depletion. Clin Diab. 2015. 33:90–5.

[40] Inose R, Takahashi K, Nishikawa T, Nagayama K. Analysis of factors influencing the development of hypomagnesemia in patients receiving cetuximab therapy for head and neck cancer. Yakugaku Zasshi. 2015. 135:1403–7.

[41] Begley J, Smith T, Barnett K, Strike P, Azim A, Spake C, Richardson T. Proton pump inhibitor associated hypomagnasaemia – a cause for concern? Br J Clin Pharmacol. 2015. DOI:10.1111/bcp.12846

[42] Hong JS, Wu LH, Su L, Zhang HR, Lv WL, Zhang WJ, Tian J. Effect of chemoradiotherapy on nutrition status of patients with nasopharyngeal cancer. Nutr Cancer. 2015. 28:1–7.

[43] Kouhen F, Afif M, Benhmidou N, El Majjaoui S, Elkacemi H, Kebdani T, Benjaafar N. What nutritional management in patients with head and neck cancers undergoing radiotherapy? An overview. Bull Cancer. 2015. 102:874–9.

[44] Håkonsen SJ, Pederson PU, Bath-Hextall F, Kirkpatrick P. Diagnostic test accuracy of nutritional tools used to identify undernutrition in patients with colorectal cancer: a systematic review. JBI Database System Rev Implement Rep. 2015. 13:141–87.

[45] Dietitian/Nutritionists from the Nutrition Education Materials Online, team Validated Malnutrition Screening and Assessment Tools: Comparison Guide Queensland Health, The Queensland Government of Australia, 2014. https://www.health.qld.gov.au/nutrition/resources/hphe_scrn_tools.pdf [Accessed: 2016-01-30].

[46] von Bokhorst-de van der Schueren M, Guiatoli APR, et al. Nutrition screening tools: does one size fit all? A systematic review of screening tools for the hospital setting. Clin Nutr. 2014. 33:39–58. DOI:10.1016/j.clnu.2013.04.008

[47] Detsky AS, McLaughlin JR, Baker JP, Johnston N, Whittaker S, Mendelson RA, Jeejeebhoy KN. What is subjective global assessment of nutritional status? J Parenter Enteral Nutr. 1987. 11:8–13. DOI:10.1177/014860701100108

[48] Thoresen L, Fjeldstad I, Krogstad K, Kaasa S, Falkmer UG. Nutritional status of patients with advanced cancer: the value of using the subjective global assessment of nutritional status as a screening tool. Pall Med. 2002. 16:33–42. DOI:10.1191/0269216302pm486oa

[49] Cooper BA, Bartlett LH, Aslani A, Allen BJ, Ibels LS, Pollock CA. Validity of subjective global assessment as a nutritional marker in end-stage renal disease.Am J Kidney Dis. 2001. 40:126–32. DOI:10.1053/ajkd.2002.33921

[50] Duerksen DR, Yeo TA, Siemens JL, O'Connor MP. The validity and reproducibility of clinical assessment of nutritional status in the elderly. Nutrition. 2000. 16:760–4. DOI: 10.1016/SO899-907(00)00398-1

[51] Christensson L, Mitra U, Anna-Christina E. Evaluation of nutritional assessment techniques in elderly people newly admitted to municipal care. Eur J Clin Nutr. 2002. 56:810–818. DOI:10.1038/sj.ejcn.1601394

[52] Sacks GS, Dearman K, Replogle WH, Cora VL, Meeks M, Canada T. Use of subjective global assessment to identify nutrition associated complications and death in geriatric long term care facility residents. J Am CollNutr. 2000. 19:570–7.DOI: 10.1080/07315724.2000.10718954

[53] Persson MD, Brismar KE, Katzarski KS, Nordenstrom J, Cederholm TE. Nutritional status using mini nutritional assessment and subjective global assessment predict mortality in geriatric patients. J Am Geriatr Soc. 2002. 50:1996–2002. DOI:10.1046/j. 1532-5415.2002.50611.x

[54] Ottery F. Patient-generated subjective global assessment. In: McCallum PD (ed.), The Clinical Guide to Oncology Nutrition. Chicago: American Dietetic Association. 2005.

[55] Isenring E, Bauer J, Capra S. Use of the scored patient-generated subjective global assessment (PG-SGA) as a nutrition assessment tool in patients with cancer. Eur J Clin Nutr. 2002. 56:779–85. DOI:10.1038/sj.ejcn.1601552

[56] Desbrow B, Bauer J, Blum C, Kandasamy A, McDonald A, Montgomery K. Assessment of nutritional status in hemodialysis patients using patient-generated subjective global assessment. J Renal Nutr. 2005. 15:211-6. DOI:10.1053/j.jrn.2004.10.005

[57] Martineau J, Bauer JD, Isenring E, Cohen S. Malnutrition determined by the patient generated subjective global assessment is associated with poor outcomes in acute stroke patients. Clin Nutr. 2005. 24:1073–7. DOI:10.1016/j.cln.2005.08.010

[58] Guigoz Y, Vellas B, GarryPJ. Mini nutritional assessment: a practical assessment tool for grading the nutritional state of elderly patients facts. In: Vellas BJ, Guigoz Y, Garry PJ, Albarede JL (eds.), Research in Gerontology. 1994. p. 15–59. ISBN: 2-909342-46-B

[59] Ferguson M, Capra S, Bauer J, Banks M. Development of a valid and reliable malnutrition screening tool for adult acute hospital patients. Nutrition. 1999. 15:458–64. DOI:10.1016/S0899-9007(99)00084-2

[60] Isenring E, Cross G, Daniels L, Kellett E, Koczwara B. Validity of the malnutrition screening tool as an effective predictor of nutritional risk in oncology outpatients receiving chemotherapy. Support Care Cancer. 2006. 14(11):1152–1156. DOI:10.1007/ s00520-006-0070-5

[61] Isenring E, Bauer JD, Banks D, MGaskill D. The malnutrition screening tool is a useful tool for identifying malnutrition risk in residential aged care. J Hum Nutr Diet. 2009. 22(6):545–50. DOI:10.1111/j.1365-277X.2009.01008.x

[62] Rubenstein LZ, Harker JO, Salva A, Guigoz Y, Vellas B. Screening for undernutrition in geriatric practice: developing the short-form Mini-Nutritional Assessment (MNA-SF). J GerontolA Biol Sci Med Sci. 2001. 56:M366–72. DOI:10.1093/gerona/56.6.M366

[63] Young A, Kidston S, Banks MD, Mudge AM, Isenring EA. Malnutrition screening tools: comparison against two validated nutrition assessment methods in older medical inpatients. Nutrition. 2013. 29:101–6. DOI:10.1016/j.nut.2012.04.007

[64] The 'MUST' Explanatory Booklet. A Guide to the 'Malnutrition Universal Screening Tool' ('MUST') for Adults in BAPEN, (BAPEN). Todorovic V. 2003.

[65] King CL, Elia M, Stroud MA, Stratton R. The predictive validity of the malnutrition screening tool ('MUST') with regard to morality and length of stay in elderly patients. Clin Nutr. 2003. 22:S4.

[66] Stratton R, Longmore D, Elia M. Concurrent validity of a newly developed malnutrition universal screening tool (MUST). Clin Nutr. 2003. 22:S10.

[67] Kondrup J, Rasmussen HH, Hamberg O, Stanga Z. Nutritional risk screening (NRS 2002): a new method based on an analysis of controlled clinical trials. Clin. Nutr. 2003. 22:321–36. DOI:10.1016/S0261-5614(02)00214-5

[68] Torre LA, Bray F., Siegel RL, Ferlay J, Lortet-Tieulent J, Jemal A. Global cancer statistics. CA Cancer J Clin. 2012. 65:87–108. DOI:10.3322/caac.21262

[69] Jolliet P, Simon N, Baree J, Pons JY, Boukef M, Paniel BJ, Tillement JP. Plasma coenzyme Q10 concentrations in breast cancer: prognosis and therapeutic consequences. Int J Clin Pharmacol Ther. 1998. 36(9):506–9.

[70] Macis D, Maisonneuve P, Johansson H, Bonanni B, Botteri E, Iodice S, Santillo B, Penco S, Gucciardo G, D'Aiuto G, Rosselli Del Turco M, Amadori M, Costa A, Decensi A. Methylenetetrahydrofolate reductase (MTHFR) and breast cancer risk: a nested-case-control study and a pooled meta-analysis. Breast Cancer Res Treat. 2007. 106(2):263–71. DOI:10.1007/s10549-006-9491-6

[71] Lee E, Levine E, Franco VI, Allen GO, Gong F, Zhang Y, Hu JJ. Combined genetic and nutritional risk models of triple negative breast cancer. Nutr Cancer. 2014. 66(6):955–63. DOI:10.1080/01635581.2014.932397

[72] Aceves C, Anguiano B, Delgado G. Is iodine a gatekeeper of the integrity of the mammary gland? J Mammary Gland Biol Neoplasia. 2005. 10(2):189–96. DOI:10.1007/s10911-005-5401-5

[73] Tseng FY, Lin W, Li CI, Li TC, Lin CC, Huang KC. Subclinical hypothyroidism is associated with increased risk for cancer mortality in adult Taiwanese-a 10 years population-based cohort. PLoS One. 2015. 10(4):e0122955. DOI:10.1371/journal.pone.0122955

[74] Babaknejad N, Sayehmire F, Sayehmiri K, Rahimifar P, Bahrami S, Delpesheh A, Hemati F, Alizadeh S. The relationship between selenium levels and breast cancer: a systematic review and meta-analysis. Biol Trace Elem Res. 2014. 159(1–3):1–7. DOI:10.1007/s12011-014-9998-3

[75] Aguirre M, Manzano N, Salas Y, Angel M, Díaz-Couselo FA, Zylberman M. Vitamin D deficiency in patients admitted to the general ward with breast, lung, and colorectal cancer in Buenos Aires, Argentina. Arch Osteoporos. 2016. 11(1):4. DOI:10.1007/s11657-015-0256-x

[76] Imtiaz S, Siddiqui N. Vitamin-D status at breast cancer diagnosis: correlation with social and environmental factors and dietary intake. J Ayub Med Coll Abbottabad. 2014. 26(2):186–90.

[77] Deschasaux M, Souberbielle JC, Latino-Martel P, Sutton A, Charnaux N, Druesne-Pecollo N, Galan P, Hercberg S, Le Clerc S, Kesse-Guyot E, Ezzedine K, Touvier M. Weight status and alcohol intake modify the association between vitamin D and breast cancer risk. J Nutr. 2016. DOI:10.3945/jn.115.221481

[78] Hurst R, Hooper L, Norat T, Lau R, Aune D, Greenwood DC, Vieira R, Collings R, Harvey LJ, Sterne JA, Beynon R, Savović J, Fairweather-Tait SJ. Selenium and prostate cancer: systematic review and meta-analysis. Am J Clin Nutr. 2012. 96(1):111–22. DOI:10.3945/ajcn.111.033373

[79] Deschasaux M, Souberbielle JC, Latino-Martel P, Sutton A, Charnaux N, Druesne-Pecollo N, Galan P, Hercberg S, Le Clerc S, Kesse-Guyot E, Ezzedine K, Touvier M. A prospective study of plasma 25-hydroxyvitamin D concentration and prostate cancer risk. Br J Nutr. 2016. 115(2):305–14. DOI:10.1017/S0007114515004353

[80] Chandler PD, GIovannucci EL, Scott JB, Bennett GG, Ng K, Chan AT, Hollis BW, Emmons KM, Fuchs CS, Drake BF. Null association between vitamin D and PSA levels among black men in a vitamin D supplementation trial. Cancer Epidemiol Biomarkers Prev. 2014. 23(9):1944–7. DOI:10.1158/1055-9965.EPI-14-0522

[81] Feldman D, Krishnan AV, Swami S, Giovannucci E, Feldman BJ. The role of vitamin D in reducing cancer risk and progression. Nat Rev Cancer. 2014. 14(5):342–57. DOI:10.1038/nrc3691

[82] Adaramoye OA, Akinloye O, Olatunji IK. Trace elements and vitamin E status in Nigerian patients with prostate cancer. Afr Health Sci. 2010. 10(1):2–8.

[83] Yan M, Song Y, Wong CP, Hardin K, Ho E. Zinc deficiency alters DNA damage response genes in normal human prostate epithelial cells. J Nutr. 2008. 138(4):667–73.

[84] Han CT, Schoene NW, Lei KY. Influence of zinc deficiency on Akt-Mdm2-p53 and Akt-p21 signaling axes in normal and malignant human prostate cells. Am J Physiol Cell Physiol. 2009. 297(5):C1188–99. DOI:10.1152/ajpcell.00042.2009

[85] Keane MG, Johnson GJ. Early diagnosis improves survival in colorectal cancer. Practitioner. 2012. 256(1753):15–8.

[86] Abbastabar H, Roustazadeh A, Alizadeh A, Hamidifard P, Valipour M, Valipour AA. Relationships of colorectal cancer with dietary factors and public health indicators: an ecological study. Asian Pac J Cancer Prev. 2015. 16(9):3991–5.

[87] Lippi G, Mattiuzzi C, Cervellin G. Meat consumption and cancer risk: a critical review of published meta-analyses. Crit Rev Oncol Hematol. 2016. 97:1–14. DOI:10.1016/j.critrevonic.2015.11.008

[88] Cho E, Zhang X, Townsend MK, Selhub J, Paul L, Rosner B, Fuchs CS, Willett WC, Giovannucci EL. Unmetabolized folic acid in prediagnostic plasma and the risk of colorectal cancer. J Natl Cancer Inst. 2015. 107(12):djv260. DOI:10.1093/jnci/djv260.

[89] Barrett CW, Singh K, Motley AK, Lintel MK, Matafonova E, Bradley AM, Ning W, Poindexter SV, Parang B, Reddy VK, Chaturvedi R, Fingleton BM, Washington MK, Wilson KT, Davies SS, Hill KE, Burk RF, Williams CS. Dietary selenium deficiency exacerbates DSS-induced epithelial injury and AOM/DSS-induced tumorigenesis. PLoS One. 2013. 8(7):e67845. DOI:10.1371/journal.pone.0067845

[90] Meeker S, Seamons A, Maggio-Price L, Paik J. Protective links between vitamin D, inflammatory bowel disease and colon cancer. World J Gastroenterol. 2016. 22(3):933–48. DOI:10.3748/wig.v22.i3.933

[91] Di Rosa M, Malaguarnera M., Zanghì A, Passaniti A, Malaguarnera L. Vitamin D3 insufficiency and colorectal cancer. Crit Rev Oncol Hematol. 2013. 88(3):564–612. DOI: 10.1016/j.critrevonc.2013.07.016

[92] Epplein M, Burk RF, Cai Q, Hargreaves MK, Blot WJ. A prospective study of plasma Selenoprotein P and lung cancer risk among low-income adults. Cancer Epidemiol Biomarkers Prev. 2014. 23(7):1238–44. DOI:10.1158/1055-9965.EPI-13-1308

[93] Xue Y, Harris E, Wang W, Baybutt RC. Vitamin A depletion induced by cigarette smoke is associated with an increase in lung cancer-related markers in rats. J Biomed Sci. 2015. 14:22–84. DOI:10.1186/s12929-015-0189-0

[94] Ho E, Courtemanche C, Ames BN. Zinc deficiency induces oxidative DNA damage and increases p53 expression in human lung fibroblasts. J Nutr. 2003. 133(8):2543–8.

[95] Khan NF, Rose PW, Evans J. Defining cancer survivorship: a more transparent approach is needed. J Cancer Survivorship. 2012. 6(1):33–6. DOI:10.1007/s11764-011-0194-6.

[96] American Cancer Society. Cancer Treatment and Survivorship Facts & Figures 2014-2015. Atlanta: American Cancer Society. 2014.

[97] McCabe MS, Bhatia S, Oeffinger KC, Reaman GH, Tyne C, Wollins DS, Hudson MM. American Society of Clinical Oncology statement: achieving high-quality cancer survivorship care. J Clin Oncol. 2013. 31(5):631–640. DOI:10.1200/JCO.2012.46.6854